A POCKET BOOK OF
MARINE ENGINEERING

Questions & Answers

First edition published in 2009
Reprinted 2013
Reprinted 2014
Reprinted 2015

ISBN: 978-1-90533-180-2

© Witherby Publishing Group Ltd, 2009-2014

British Library Cataloguing in Publication Data
A catalogue record for this book is available from the British Library.

All rights reserved. No part of this publication may be reproduced, stored in a retrieval system, or transmitted in any form or by any means, electronic, mechanical, photocopying, recording or otherwise, without the prior permission of the publishers.

While the advice given in this book 'A Pocket Book of Marine Engineering Questions & Answers' has been developed using the best information currently available, it is intended purely as guidance to be used at the user's own risk. Witherby Seamanship International Ltd accepts no responsibility for the accuracy of any information or advice given in the document or any omission from the document or for any consequence whatsoever resulting directly or indirectly from compliance with or adoption of guidance contained in the document even if caused by failure to exercise reasonable care.

This publication has been prepared to deal with the subject of 'A Pocket Book of Marine Engineering Questions & Answers'. This should not however, be taken to mean that this publication deals comprehensively with all of the issues that will need to be addressed or even, where a particular issue is addressed, that this publication sets out the only definitive view for all situations.

The opinions expressed are those of the author only and are not necessarily to be taken as the policies or views of any organisation with which he has any connection.

Printed and bound in Great Britain by Charlesworth Press, Wakefield

Published by
Witherby Publishing Group Ltd
4 Dunlop Square, Livingston,
Edinburgh, EH54 8SB,
Scotland, UK

Tel No: +44(0)1506 463 227
Fax No: +44(0)1506 468 999

Email: info@emailws.com
Web: www.witherbys.com

CONTENTS

QUESTION

1 **Firefighting** 3
    1.1   Questions - Fire Prevention and Sources of Ignition on Ships.... 3
    1.2   Questions - Inert Gas and Extinguishers 3
    1.3   Questions - CO_2, Sprinkler Systems and Detection 4
    1.4   Questions - Classification Regulations 5

2 **Safety** 7
    2.1   Questions - Enclosed Spaces and Entry 7
    2.2   Questions - Lifting Plant and Welding 8
    2.3   Questions - Pollution and Harmful Substances 8

3 **Materials** 11

4 **Auxiliary Machinery** 15
    4.1   Questions - Heat Exchangers 15
    4.2   Questions - Pumps 15
    4.3   Questions - Filters 16
    4.4   Questions - Stabilisers 16
    4.5   Questions - Refrigeration 17

5 **Engine Room Systems** 19
    5.1   Questions - Bilge And Ballast System 19
    5.2   Questions - Fresh Water and Seawater Cooling Systems 19
    5.3   Questions - Fuel Oil System 20
    5.4   Questions - Engine Room Systems (General) 21

6 **Ship Construction** 25
    6.1   Questions - Ship Construction 25

7 **Steering Gear** 27
    7.1   Questions - Steering Gear 27

8 **Shafting** 29
    8.1   Questions - Shafting 29

9 Electrical... 31
   9.1   Questions - Switchboards, Circuit breakers and Switchboard Instrumentation.. 31
   9.2   Questions - Parallel Operation of Generators............................ 31
   9.3   Questions - Motor Starters 32
   9.4   Questions - Shore Supply.. 32
   9.5   Questions - Electrical (General)... 33

10 Electrical Protection 35
   10.1   Questions - Electrical Protection (General)............................... 35
   10.2   Questions - Fault Protection 35
   10.3   Questions - Circuit Breaker... 35
   10.4   Questions - Overcurrent Protection... 36
   10.5   Questions - Fuses... 36
   10.6   Questions - Protection Discrimination....................................... 36
   10.7   Questions - Loss of Excitation 37
   10.8   Questions - Undervoltage Protection.. 37
   10.9   Questions - Reverse Power Protection..................................... 37
   10.10 Questions - Motor Overload Protection 38
   10.11 Questions - Earth Faults 38

11 Electrical Safety 39
   11.1   Questions - General Electrical Maintenance............................. 39
   11.2   Questions - Switchboard Maintenance 39
   11.3   Questions - Motor Maintenance 40
   11.4   Questions - Motor Control Gear Maintenance 41
   11.5   Questions - Maintenance of Lighting 41
   11.6   Questions - Electric Shock.. 42
   11.7   Questions - Circuit Safety Devices.. 42
   11.8   Questions - Earth Safety Devices 43
   11.9   Questions - Welding 43
   11.10 Questions - Static Electricity and Interference.......................... 43
   11.11 Questions - Emergency Power Supplies 44

12 Instrumentation and Control 47
   12.1   Questions - Control (General)... 47

12.2	Questions - Control Systems	47
12.3	Questions - Measuring Instruments	48
12.4	Questions - Automatic Control Theory	50
12.5	Questions - Types of Control Action	52
12.6	Questions - Automatic Controllers	52
12.7	Questions - Control System Components	54
12.8	Questions - Supply Systems	56
12.9	Questions - Practical Control Systems	56
12.10	Questions - Alarm Indication Systems	58
12.11	Questions - Operation And Maintenance	59

13 Diesel Engine Theory & Structure 63

13.1	Questions - Engine Types	63
13.2	Questions - Engine Construction	64
13.3	Questions - The Crankshaft	64
13.4	Questions - Connecting Rods and Crossheads	65
13.5	Questions - Pistons	66
13.6	Questions - Cylinder Diaphragm and Piston Rod Gland	68
13.7	Questions - Cylinder Liner	68
13.8	Questions - Cylinder Cover	70
13.9	Questions - Camshafts and Valve Gear	70
13.10	Questions - Turbocharge Systems	72

14 Fuel Systems 75

14.1	Questions - Fuel Oil	75
14.2	Questions - Supply of Fuel Oil	77
14.3	Questions - Fuel Pumps	78
14.4	Questions - Fuel Injectors	79

15 Combustion 81

15.1	Questions - Fuel and Fuel Burning (Diesel Engines)	81

16 Lubrication 83

16.1	Questions - The lubricating oil system	83
16.2	Questions - Lubricating Oils (1)	83
16.3	Questions - Lubrication Fundamentals	85
16.4	Questions - Lubricating Oils (2)	86

	16.5	Questions - Shipboard Lubricating Oil Tests	89
	16.6	Questions - Grease	89
	16.7	Questions - Oil Drain Tank	90
	16.8	Questions - Pre-Cleaning and Corrosion Protection	91
	16.9	Questions - Rotary Displacement Pumps	92
	16.10	Questions - Coolers	92
	16.11	Questions - Oil Maintenance	94

17 Starting & Manoeuvring Safety ... 97
- 17.1 Questions - Starting Air ... 97
- 17.2 Questions - Manoeuvring/Direct Reversing Engines ... 98
- 17.3 Questions - Safety Systems (Crankcase) ... 100

18 Boilers ... 103
- 18.1 Questions - Boiler Types ... 103
- 18.2 Questions - Evaporators ... 103
- 18.3 Questions - Boiler Mountings ... 105
- 18.4 Questions - Water Gauges ... 106
- 18.5 Questions - Boiler Operation ... 108
- 18.6 Questions - Tube Failures ... 110

ANSWERS

1 Firefighting ... 113
- 1.1 Answers - Fire Prevention and Sources of Ignition on Ships ... 113
- 1.2 Answers - Inert Gas and Extinguishers ... 115
- 1.3 Answers - CO_2, Sprinkler Systems and Detection ... 119
- 1.4 Answers - Classification Regulations ... 123

2 Safety ... 129
- 2.1 Answers - Enclosed Spaces and Entry ... 129
- 2.2 Answers - Lifting Plant and Welding ... 135
- 2.3 Answers - Pollution and Harmful Substances ... 136

3 Materials ... 143

4 Auxiliary Machinery **151**
- 4.1 Answers - Heat Exchangers 151
- 4.2 Answers - Pumps 157
- 4.3 Answers - Filters... 163
- 4.4 Answers - Stabilisers.. 167
- 4.5 Answers - Refrigeration 170

5 Engine Room Systems.. **173**
- 5.1 Answers - Bilge and Ballast System..................................... 173
- 5.2 Answers - Fresh Water and Sea Water Cooling Systems 176
- 5.3 Answers - Fuel Oil System.. 179
- 5.4 Answers - Engine Room Systems (General)............................ 184

6 Ship Construction **191**
- 6.1 Answers - Ship Construction.. 191

7 Steering Gear **199**
- 7.1 Answers - Steering Gear 199

8 Shafting.. **207**
- 8.1 Answers - Shafting 207

9 Electrical... **213**
- 9.1 Answers - Switchboards, Circuit breakers, Switchboard Instrumentation 213
- 9.2 Answers - Parallel Operation of Generators 214
- 9.3 Answers - Motor Starters... 214
- 9.4 Answers - Shore Supply 215
- 9.5 Answers - Electrical (General)... 216

10 Electrical Protection **221**
- 10.1 Answers - Electrical Protection (General)............................... 221
- 10.2 Answers - Fault Protection... 221
- 10.3 Answers - Circuit Breaker 221
- 10.4 Answers - Overcurrent Protection 222
- 10.5 Answers - Fuses 224
- 10.6 Answers - Protection Discrimination 224

10.7	Answers - Loss of Excitation	225
10.8	Answers - Undervoltage Protection	226
10.9	Answers - Reverse Power Protection	226
10.10	Answers - Motor Overload Protection	227
10.11	Answers - Earth Faults	228

11 Electrical Safety 231

11.1	Answers - General Electrical Maintenance	231
11.2	Answers - Switchboard Maintenance	232
11.3	Answers - Motor Maintenance	233
11.4	Answers - Motor Control Gear Maintenance	235
11.5	Answers - Maintenance of Lighting	236
11.6	Answers - Electric Shock	236
11.7	Answers - Circuit Safety Devices	237
11.8	Answers - Earth Safety Devices	237
11.9	Answers - Welding	238
11.10	Answers - Static Electricity and Interference	238
11.11	Answers - Emergency Power Supplies	239

12 Instrumentation and Control 243

12.1	Answers - Control (General)	243
12.2	Answers - Control Systems	243
12.3	Answers - Measuring Instruments	245
12.4	Answers - Automatic Control Theory	252
12.5	Answers - Types of Control Action	255
12.6	Answers - Automatic Controllers	255
12.7	Answers - Control System Components	258
12.8	Answers - Supply Systems	260
12.9	Answers - Practical Control Systems	262
12.10	Answers - Alarm Indication Systems	265
12.11	Answers - Operation and Maintenance	267

13 Diesel Engine Theory & Structure 271

13.1	Answers - Engine Types	271
13.2	Answers - Engine Construction	272
13.3	Answers - The Crankshaft	274

    13.4   Answers - Connecting Rods and Crossheads 276
    13.5   Answers - Pistons.. 277
    13.6   Answers - Cylinder Diaphragm and Piston Rod Gland............ 280
    13.7   Answers - Cylinder Liner 281
    13.8   Answers - Cylinder Cover.. 283
    13.9   Answers - Camshafts and Valve Gear 284
    13.10 Answers - Turbocharge Systems.. 287

14 Fuel Systems...**291**
    14.1   Answers - Fuel Oil 291
    14.2   Answers - Supply of Fuel Oil 294
    14.3   Answers - Fuel Pumps... 297
    14.4   Answers - Fuel Injectors 298

15 Combustion...**301**
    15.1   Answers - Fuel and Fuel Burning (Diesel Engines)................. 301

16 Lubrication **305**
    16.1   Answers - The Lubricating Oil System..................................... 305
    16.2   Answers - Lubricating Oils (1).. 306
    16.3   Answers - Lubrication Fundamentals....................................... 308
    16.4   Answers - Lubricating Oils (2).. 310
    16.5   Answers - Shipboard Lubricating Oil Tests.............................. 314
    16.6   Answers - Grease... 315
    16.7   Answers - Oil Drain Tank 316
    16.8   Answers - Pre-Cleaning and Corrosion Protection 317
    16.9   Answers - Rotary Displacement Pumps.................................. 319
    16.10 Answers - Coolers 320
    16.11 Answers - Oil Maintenance.. 322

17 Starting & Manoeuvring Safety **327**
    17.1   Answers - Starting Air.. 327
    17.2   Answers - Manoeuvring / Direct Reversing Engines 329
    17.3   Answers - Safety Systems (Crankcase)................................... 332

18 Boilers **337**
- 18.1 Answers - Boiler Types 337
- 18.2 Answers - Evaporators 338
- 18.3 Answers - Boiler Mountings 340
- 18.4 Answers - Water Gauges 343
- 18.5 Answers - Boiler Operation 343
- 18.6 Answers - Tube Failures 349

QUESTIONS

1 FIREFIGHTING

1.1 Questions – Fire Prevention and Sources of Ignition on Ships

1. Give some examples from your own experience of the ways in which carelessly carried out maintenance may cause a fire?

2. What are the safeguards against common causes of electrical fires?

3. What causes spontaneous combustion?

4. a) State where information can be obtained with regard to the safe carriage of hazardous substances as cargoes.
   b) For a hazardous substance of your choice discuss EACH of the following:
      i. Storage and transport
      ii. Properties
      iii. Fire-fighting techniques
      iv. Medical effects and treatment after physical contact with the substance. (If your ship does not carry a hazardous cargo, choose a substance onboard which would be considered hazardous and answer the question accordingly.)

5. Describe an example onboard your ship of a welding operation and explain in detail the precautions taken against fire and explosion.

6. Describe the fuel tank overflow arrangements on your vessel with particular reference to location and safety design features.

1.2 Questions – Inert Gas and Extinguishers

1. Sketch and describe an inert gas installation for an oil tanker and detail what safety features and alarms are fitted to the system.

Engineering Pocket Book – Questions

    2. Which fire-fighting media would you use on the following types of fire, explain why you chose that particular media and detail the fire-fighting procedure:

        a. A small oil fire in the machinery space.
        b. A bedding fire in the accommodation.
        c. A galley fryer has been left switched on, the thermostat has failed and the oil has burst into flames.

    3. Sketch and describe the following portable fire extinguishers:

        a. Portable foam.
        b. Dry powder.

    4. Define the term 'flammable limits'.

1.3. Questions – CO_2, Sprinkler Systems and Detection

    1. What are the advantages and disadvantages of each of the following fixed fire fighting systems:

        a. Carbon Dioxide flooding system?
        b. Water spray systems?
        c. High expansion foam system?

    2. Describe, with the aid of sketches, the fixed fire-fighting system for the machinery spaces onboard your vessel.

    3. Describe the type, location and operation of the emergency fire pump installed on your vessel. Explain how this pump fulfils the Classification requirements.

    4. Write a short note on each of the following fire detection methods:

        a. Fire patrols.
        b. Heat sensors.
        c. Infrared flame sensors.
        d. Electric photo cell smoke detectors.

    5. Explain how the above detectors are tested in situ.

1.4 Questions – Classification Regulations

1. With reference to a water sprinkler fixed fire-fighting system suitable for accommodation spaces:

   (a) Sketch such a system and explain its method of operation.
   (b) Discuss the advantages and disadvantages of such a system.

2. (a) Sketch a bulk carbon dioxide fixed fire-fighting system suitable for machinery spaces and for the cargo spaces of a dry cargo ship.
   (b) i) Describe the system you have sketched and its method of operation.
   ii) State the advantage this system has over the multi-bottle carbon dioxide system.

3. (a) Write a list of the various means by which explosive gas may be ignited inadvertently by electrical machinery.
   (b) State how the probability of the ignition of explosive gas by electrical machinery may be reduced.

4. Shore would it be used ? Make a sketch of the international shore connection and indicate its dimensions.

2 SAFETY

2.1 Questions – Enclosed Spaces and Entry

1. State why oxygen deficiency may occur in certain spaces onboard ship.

2. Describe the precautions taken before entry to any recently opened space.

3. a) Describe, with the aid of a sketch, any type of self-contained breathing apparatus.
   b) State why a bottle is checked before use and how warning is given that the bottle contents are nearly exhausted.

4. List the gases which might be present in:

   a. Permanent ballast tanks.
   b. Cargo Tank.
   c. Pump room.
   d. Fresh water tank.

5. a) Describe a 'permit to work' form.
   b) When is a 'permit to work' form required?

6. Detail the toxicity of oil cargoes or fuels.

7. Starting from a tank which has contained oil cargo and is fitted with an inert gas system, using the graph of air/fuel which indicates the flammable range, explain the safest method of gas-freeing the tank and preparing the tank for entry. Explain the use of any gas analysis instruments used.

2.2 Questions – Lifting Plant and Welding

1. In the context of a ship's crane, explain the meaning of the following:

   a) Raising and lowering.
   b) Slewing.
   c) Luffing.

2. List the safety precautions necessary while working aloft. Give an example of work carried out on your vessel which required working aloft and detail the procedures.

3. What precautions should the Duty Engineer take while working in the vicinity of the funnel?

2.3 Questions – Pollution and Harmful Substances

1. What are the dangers of asbestos dust? If it is essential to carry out repairs liable to create asbestos dust, what precaution would you take?

2. With reference to static oil/water separators:

   a. State the forces which contribute towards the total force available for oil and water separation.
   b. With the aid of a sketch, describe the passage of oil/water flowing through the separator, discussing how the separator's design is adapted to utilise the forces you have stated in part a).
   c. State how oil density and temperature relate to ease of separation.

3. Describe the following with the aid of a sketch a shipboard sewage treatment plant.

   - Explain what is meant by Colliform Count
   - Explain the term Biological Oxygen Demand

- State with reasons FOUR operational circumstances that may adversely affect the quality of the output of the system.

4. What are the dangers of producing domestic water close to the shoreline? Explain how onboard your vessel domestic water is produced and detail the treatment carried out to the water.

Engineering Pocket Book – Questions

3 MATERIALS

1. Why is phosphorus undesirable in steel?

2. Above what percentage of manganese does steel become an alloy?

3. What is the benefit of adding manganese to steel?

4. How does temperature affect grain boundaries?

5. Upon what factors does the recrystallisation of metal depend?

6. What is the difference between hot working and cold working and what differences do these produce in the metal?

7. State the percentages of carbon in the following types of steel:
   a) Dead mild steel
   b) mild steel
   c) medium carbon steel
   d) high carbon steel.

8. Give advantages and disadvantages for the use of each of the above steels.

9. Which steel is the most generally used and why?

10. What is the purpose of heat treating steels?

11. Give three types of heat treatment and the purpose of this heat treatment.

12. Describe the process of nitriding and give two examples of where it could be used.

13. State why alloy steel is necessary and give reasons.

14. When would austenitic stainless steels be used?

15. What are the benefits of using titanium alloys?

Engineering Pocket Book – Questions

16. What are the nimonic series of metals?

17. What is stellite and where would it be used?

18. Describe four alloying elements, stating the percentages used and the properties given to steel.

19. What is fatigue in relation to metals?

20. What are the factors that influence the fatigue strength of a material?

21. Give a brief description of creep?

22. State the factors that influence creep.

23. What are the constituents of cast iron?

24. State four properties of cast iron.

25. Why is white cast iron undesirable?

26. What factors affect the form taken by carbon in cast iron?

27. What is spheroidal graphite cast iron?

28. Where would copper be used and why?

29. Describe dezincification of brass.

30. How is dezincification of brass prevented?

31. What properties should bearing materials have?

32. How are sintered bearings made?

33. Give a brief description of electrochemical corrosion.

34. Where is the corrosion most severe?

35. What is the difference between the anode and the cathode?

36. What is pitting?

37. Describe TIG welding.

38. Describe uses for compressed asbestos fibre jointing.

39. What are the advantages of PTFE as a jointing material?

40. State briefly the limitations of PTFE.

41. What are the factors that affect sealing characteristics?

Engineering Pocket Book – Questions

4 AUXILIARY MACHINERY

4.1 Questions – Heat Exchangers

1. Which points need to be considered in the design/selection of a heat exchanger?

2. Define streamline and turbulent flow. What is the importance of streamline and turbulent flow to the design/selection of heat exchangers?

3. Sketch and describe a shell and tube type heat exchanger.

4. Sketch and describe a plate type heat exchanger.

5. What are the principal advantages of plate type heat exchangers?

6. What are the functions of a condenser?

7. What are the two principal types of condenser?

8. Sketch and describe an 'end on' or axial exhaust regenerative condenser.

9. List the materials used in the construction of a condenser

10. How are tubes fixed in a condenser?

11. How is a defective condenser tube located?

4.2 Questions – Pumps

1. Describe the passage of water through a centrifugal pump.

2. Sketch and describe a centrifugal pump showing the impeller and volute casing.

3. What are positive displacement pumps?

Engineering Pocket Book – Questions

4. What are dynamic pressure pumps?

5. Sketch an axial flow pump.

6. Sketch and describe a gear pump.

7. Sketch and describe a piston pump.

8. What type of pump would you choose to use as an emergency bilge pump?

9. What is cavitation? When and where might it occur?

4.3 Questions – Filters

1. Sketch and describe an auto-klean strainer.

2. Sketch and describe a streamline lubricating oil filter.

3. What is coalescing action?

4. What is the size limit of particles which can be removed using an auto-klean strainer?

5. Where are pressure gauges fitted on the auto-klean filter?

6. How is a lubricating oil coalescer-filter cleaned?

7. Sketch the back flush cleaning arrangement of an automatic oil filter module.

4.4 Questions – Stabilisers

1. Where are stabilising fins fitted?

2. What is the purpose of the tail flap?

3. What is the angle of operation of the main fin?

4. What is the angle of operation of the tail flap?

5. What is the hydraulic pressure required to extend and house the fins?

6. What is the hydraulic pressure required to alter fin angle?

7. What instrument is used to sense the motion of the ship and then send a signal to the fin controls?

8. What panel is fitted to the bridge controls to monitor fin position?

4.5 Questions – Refrigeration

1. Name the four basic components of a refrigeration system, working on the vapour compression cycle.

2. Explain the function of the compressor.

3. Explain the function of the condenser.

4. Explain the function of the expansion valve.

5. Explain the purpose of the evaporator.

6. Name three desirable properties of a refrigerant.

7. Where is the high pressure cut out fitted in the refrigeration circuit?

8. What are the effects of insufficient refrigerant in the system?

5 ENGINE ROOM SYSTEMS

5.1 Questions – Bilge and Ballast System

1. Make a diagrammatic sketch of a bilge pumping system.
2. Itemise the main components of the bilge pumping system.
3. Make a diagrammatic sketch of a ballast pumping system.
4. Itemise the main components of a ballast pumping system.
5. What type of valve will be used in the ballast system?
6. What type of valve will be used in the bilge system?
7. Sketch a bilge injection valve.
8. What is the function of a bilge injection valve?
9. What should the diameter of the bilge injection valve be in relation to the diameter of the main sea inlet?
10. What is the function of the bilge system?
11. What is the function of the ballast system?

5.2 Questions – Fresh Water and Seawater Cooling Systems

1. What are the advantages of a central cooling system?
2. List the four checking and treatment procedures used to reduce the possibility of damage to fresh water systems.
3. How may corrosion occur in a fresh water cooling system?
4. What is the purpose of the expansion tank in a fresh water cooling system?

Engineering Pocket Book – Questions

    5.   Sketch a system diagram for cooling a diesel engine which does not use a central cooling system.

    6.   When is the upper sea suction valve used?

    7.   When is the lower sea suction valve used?

    8.   Why may steam and compressed air be supplied to the seachest valve?

    9.   Sketch and describe a system which makes it possible to heat the main engines by means of cooling water from the auxiliary engines.

   10. What are the advantages of the above system?

5.3 Questions – Fuel Oil System

    1.   Sketch and describe a fuel oil system for use with a boiler.

    2.   Sketch and describe a fuel oil system suitable for use with a diesel engine.

    3.   Sketch and describe a settling tank.

    4.   What is the function of the settling tank?

    5.   Why is the temperature of fuel oil raised before burning?

    6.   How is the correct oil temperature maintained?

    7.   Two emergency valves are fitted in a boiler fuel oil system. What are their locations and functions?

    8.   What is the purpose of a recirculating valve fitted in the boiler fuel oil system?

    9.   List the safety fittings in a boiler fuel oil system.

   10. Describe how a boiler fuel oil system can be changed from automatic to manual control.

Engineering Pocket Book – Questions

11. Why is the density of fuel oil burned in a diesel engine fuel oil system important?

12. Why is the viscosity of fuel oil burned in a diesel engine fuel oil system important?

13. Why is the flash point of fuel in a fuel oil system important?

14. What solid contaminants might be found in fuel oil?

15. What liquid contaminants might be found in fuel oil?

16. Why are the two centrifuges of a diesel engine fuel oil system arranged in series?

17. List the various safety devices included in a diesel engine fuel oil system.

18. Automatic control of the heater in a diesel engine fuel oil system may be by measurement of viscosity or by measurement of temperature. Why is control by measurement of viscosity considered superior?

5.4 Questions – Engine Room Systems (General)

1. Give reasons why fuel oil pipes must be clipped and supported.

2. In large 2-stroke engines, the fuel injectors will circulate fuel during periods in which they are not actually injecting into the cylinder. What is the reason for this?

3. In a diesel engine fuel oil system, where does oil which has been recirculated return to?

4. When burning heavy fuel oil in a diesel engine, why is it necessary to reduce the high viscosity of the fuel?

5. What is the recommended standard treatment of residual fuel to be used in large engines?

6. It is advisable to store ship's bunkers from different ports in separate tanks. What difficulties may arise if this precaution is neglected?

7. In a boiler fuel oil system, the correct oil temperature is maintained by means of a thermostat placed in the outlet from the fuel oil heater. What does this thermostat control?

8. Boiler fuel oil systems use gas oil to flash up from cold. How is this achieved?

9. What is the function of the dumping valve fitted to a settling oil tank?

10. What is the function of the sludge valve or cock fitted to a settling oil tank?

11. The exhaust steam from heating coils will be led to a steam trap. What is the function of this?

12. The central cooling system is divided into low temperature and high temperature zones. Sketch and describe both the low temperature zone. and the high temperature zone.

13. In an automatic domestic water supply system, the tank containing water has an air space provided above the water. What is the function of the compressed air in this space?

14. In a domestic water purification system, why are carbonates of calcium and magnesium used in a neutraliser?

15. In the above system, why is chlorine used and in what concentrations?

16. Sketch and describe a central priming system.

17. What is the function of a central priming system?

18. State the advantages of a central priming system.

19. In a central priming system, how is water prevented from entering the vacuum tank after priming?

20. Draw the British Standard symbol for the following valves:-

- Screw down valve
- non-return valve
- safety valve
- butterfly valve
- double seated change-over valve.

21. Draw the British Standard symbol for the following appliances:-

- Suction box
- filter or strainer
- steam trap
- separator.

22. Draw the British Standard symbol for the following fittings, indicating and measuring instruments:-

- Funnel
- open scupper with closing device
- pipe flow indicator
- pressure gauge
- thermometer.

23. Why are bilge and ballast systems interconnected?

24. How are bilge and ballast systems interconnected?

6 SHIP CONSTRUCTION

6.1 Questions – Ship Construction

1. Make a sketch of a longitudinal section of a ship, identifying the principal parts.

2. Make a sketch of a transverse section of a ship, identifying the principal parts.

3. Where is the loadline disc positioned on the hull?

4. Explain the term 'panting'.

5. Explain the difference between 'strake' and 'stringer'.

6. What purpose do 'floors' serve?

7. With regard to pipework, what is the advantage of a duct keel?

8. Name two purposes for the double bottom.

9. In what areas of the ship are the number of floors increased?

10. Why are bilge keels fitted?

11. What types of steel are used in tanker construction?

12. In the construction of tankers, above what length of vessel does Lloyd's Register generally require full longitudinal framing?

13. Where are cross ties fitted in tanker construction?

14. What spaces may be accepted in lieu of cofferdams in tanker construction?

15. Why may a secondary barrier be required as a lining between the cargo container and the ship's hull in a gas carrier?

16. Describe a membrane tank in a gas carrier.

17. Name two materials used in the construction of liquefied gas tanks.

18. To what use may the 'boil off' vapour from an LNG ship's tanks be put if the ship is steam powered?

19. Where is the collision bulkhead located?

20. What safety notice should be prominently displayed at each MacGregor type rolling hatch cover?

21. Sketch a watertight door.

22. What arrangements are made to allow personnel to pass through a watertight bulkhead?

23. Name three ship Classification Societies.

7 STEERING GEAR

7.1 Questions – Steering Gear

1. Name three basic requirements for steering gears.

2. Why are jumping stops fitted?

3. In ram type hydraulic steering gear, how is side loading relieved on the cylinder glands?

4. How are ram cylinders braced against hydraulic pressure which tends to push them apart?

5. How can the rudder be locked, in an emergency, in a hydraulic steering gear?

6. How is hydraulic steering gear protected against overloading by heavy seas on the rudder?

7. In hydraulic steering gears, why should special attention be paid to high pressure oil pipes (especially after heavy weather)?

8. What rudder angle is permitted by a rotary vane steering gear with two moving and two fixed vanes?

9. Describe how the rotor is attached to the rudder in a rotary vane steering gear.

10. How are the vanes sealed in a rotary vane steering gear?

11. What prevents rotation of the stator in a rotary vane steering gear?

12. How many radial cylinders are there in a Hele-Shaw pump?

13. How is the pumping action produced in a Hele-Shaw pump?

14. How is the oil flow varied in a VSG pump?

Engineering Pocket Book – Questions

15. How is the idle pump prevented from running in the reverse direction when two pumping units are fitted?

16. What fitting restores the rudder to its required position after it has been displaced by a heavy sea?

17. Name three types of control systems for electro-hydraulic steering gears.

18. What is the working fluid in a telemotor control system?

19. What precaution must be taken when charging a hydraulic steering system, so that it will operate properly?

20. How is control effected on an electro-hydraulic steering gear with fixed capacity pumps?

21. When must the main steering and emergency steering gear be tested?

22. Who should be notified before the steering gear is operated?

23. What action will the person in Question 22 take?

24. List the steps involved in testing the steering gear?

25. What are the installation precautions taken for supply circuits from the main switchboard to the steering gear?

26. What electrical protection is given to the steering gear circuits?

8 SHAFTING

8.1 Questions – Shafting

1. Propulsion shaft bearings divide into two groups. What are these groups?

2. Do shaft bearings load vary with rpm?

3. Why is reliability particularly important in propulsion shaft bearings?

4. How are shaft support bearings lubricated?

5. What are the advantages of using roller bearings for propulsion shaft support?

6. Why are roller bearings not generally used to support the larger size of propulsion shaft?

7. Sketch a self-aligning propulsion shaft bearing.

8. What might cause a propulsion shaft to hog or sag while in use?

9. Sketch an oil lubricated stern tube.

10. Sketch a stern tube lubricated oil system.

11. How are the holding down bolts of a thrust block relieved of shear?

12. Sketch and describe a ship side valve.

13. Why are screw down non-return valves used as ship side valves?

14. What are the advantages of hydraulic-powered valved actuators?

15. How is damage to valves by powered valve actuators prevented?

16. Outline the care and maintenance of ship side valves.

9 ELECTRICAL

9.1 Questions – Switchboards, Circuit breakers and Switchboard Instrumentation

1. What is the duty of a switchboard?
2. What is the purpose of a circuit breaker?
3. What duties are springs used for in circuit breakers?
4. What current rating range is used with MCCBs?
5. What maintenance can be carried out on an MCBs?
6. What does an ammeter measure?
7. What does a wattmeter measure?
8. What is the voltage in the secondary winding of a voltage transformer?
9. When voltages are equal in frequency, where does the synchroscope pointer lie?
10. What is the maximum time a synchroscope should be on line for?

9.2 Questions – Parallel Operation of Generators

1. When must the circuit breaker be closed when paralleling?
2. When using a synchroscope, in which direction should the pointer be travelling before closing the circuit breaker?
3. When using lamps, when is the correct moment to close the circuit breaker?
4. What should be done after successful synchronisation?

5. What is varied to balance alternator load?

6. What is used to adjust the power factor?

9.3 Questions – Motor Starters

1. What is the most common type of starter at sea?

2. What is the starting current compared to full load current?

3. When connected in star, what is the voltage as a percentage of full load voltage?

4. What is the most common reduced voltage starter?

5. With a manual change-over in a star-delta starter, what is the interlock for?

6. With a star-delta starter, when is the change-over?

7. What are the voltage tappings with an auto transformer?

8. With auto transformer starters, what is used to overcome the transition switching problem?

9. What kind of starters are now becoming common at sea?

9.4 Questions – Shore Supply

1. What is a shore supply required for?

2. Can you normally parallel shore supply with the ship's alternator?

3. How do you correct phase sequence?

4. If a 60Hz ship is supplied with a 50 Hz supply, what is the reduction in voltage?

9.5 Questions – Electrical (General)

1. Name the four kinds of switchboard?
2. What are the instruments found on a switchboard?
3. What should not be used to dress contacts of a circuit breaker?
4. What kind of closing mechanisms are used in circuit breakers?
5. What can cause a circuit breaker to trip?
6. What is the typical insulation resistance of an MCB?
7. What is the current range used with MCBs?
8. What is the instrumentation required for parallel operation?
9. What does a voltmeter measure?
10. What currents are normally found in the secondary winding of a current transformer?
11. What is the most common form of ammeter or voltmeter?
12. What kind of instrument is used for a voltmeter?
13. What are the phases to which the synchronising lamps are attached?
14. What is the problem associated with paralleling ac generators?
15. What are the operations necessary for paralleling?
16. Why must the synchroscope be moving in the fast direction?
17. When using a synchroscope, when is the correct moment to close the circuit breaker?
18. What is the typical droop on a alternator governor?
19. What is used to prevent an error in synchronising?

Engineering Pocket Book – Questions

20. What is monitored by a good check synchroniser?

21. Why is it undesirable to make repeated successive starts?

22. What is the disadvantage of a direct-on-line starter?

23. What is the starting torque when motor windings are in star as a percentage of full load torque?

24. What kind of starter is used for starting large motors?

25. What is the minimum period for cooling after two consecutive starts with an auto transformer starter?

26. What is meant by intermittent duty with regard to starters?

27. 'Soft start' starters use what to control current flow during starting?

28. What are the adjustments that can be made to an electronic starter?

29. What happens when motors run at a higher frequency?

30. If a 50 Hz ship is supplied with 60 Hz, what is the increase in centrifugal loads?

31. What happens if a motor stalls?

32. Heating of a motor varies in what proportion to the current?

10 ELECTRICAL PROTECTION

10.1 Questions – Electrical Protection (General)

1. Above what voltage must special precautions be taken?

2. Why are high voltage systems more dangerous than low/medium systems?

10.2 Questions – Fault Protection

1. What is the most serious hazard at sea?

2. What is the maximum temperature allowable with rated full load current?

3. What happens to insulation at high temperatures?

10.3 Questions – Circuit Breaker

1. What are the principal factors to be considered when selecting a circuit breaker?

2. What extra factors are considered with high voltage systems?

3. What factors determine the current rating of a circuit breaker?

4. By what percentage would the rating of a circuit breaker change from its free air value due to switchboard mounting?

5. What is the primary function of a circuit breaker?

6. What are the four fault ratings of a circuit breaker?

7. What would happen if the circuit breaker was rated at less than the expected fault level?

8. What determines the size of the short circuit fault current?

Engineering Pocket Book – Questions

10.4 Questions – Overcurrent Protection

1. What is system protection discrimination?
2. What problem can occur if the full load current is greater than the reset current?
3. Why can the fault current be higher when the alternator is cold?
4. In practice, what is the time delay from 10% full load capacity?
5. What is the usual short circuit condition time delay for alternator overcurrent protection?
6. What is the common characteristic of all overcurrent relays?
7. Why is higher viscosity oil used in marine dashports?
8. How does an electronic overcurrent relay operate?
9. What kind of overcurrent protection is found in MCCBs and MCBs?
10. How is the overcurrent protection tested?

10.5 Questions – Fuses

1. What specifies the size of a fuse?
2. What kind of fuse is not recommended for use at sea?
3. What is the advantage of a fuse?
4. What must be done before replacing a fuse?
5. What does the symbol on an HRC fuse link represent?

10.6 Questions – Protection Discrimination

1. How is protection discrimination achieved?
2. What course of action needs to be taken when a generator is overloaded?

3. Apart from overload, what can initiate preference tripping?

4. How is preference tripping tested?

5. In alternators above 5000 kVa, what is required in the machine windings?

10.7 Questions – Loss of Excitation

1. What happens when there is a loss of excitation in a parallel alternator system?

2. In the event of the above situation occurring, which alternator would trip and which trip would operate?

10.8 Questions – Undervoltage Protection

1. At what value of voltage drop does the undervoltage trip operate?

2. Why are motor starters fitted with undervoltage trips?

3. What kind of relays are used for undervoltage protection?

10.9 Questions – Reverse Power Protection

1. What does a reverse power relay monitor?

2. Why is a steam turbine overspeed trip connected to the circuit breaker trip?

3. With diesel sets, what extra load occurs when one set motors?

4. What is the reverse power delay setting for steam turbine sets?

5. Why is there a delay on the relay?

6. What can be used instead of reverse power protection.

Engineering Pocket Book – Questions

10.10 Questions – Motor Overload Protection

1. What is the most common form of overload protection?
2. What is the advantage of a thermal relay over a magnetic relay?
3. What is the disadvantage of thermal relay?
4. How does a magnetic relay operate?
5. What is used to delay the operation of a magnetic relay?
6. What built-in motor overload protection can be provided?
7. What does this protect against primarily?
8. What is the advantage of the thermostat over the thermistor?
9. What features can high specification electronic relays provide?

10.11 Questions – Earth Faults

1. What happens when these is one earth fault in an earthed distribution system?
2. Which distribution system is more efficient in maintaining supply?
3. What consideration is used to choose the value of an earthing resistor?
4. What is the maximum voltage of electrical systems in tanker cargo areas?
5. What is the sign of an earth fault using earth lamps?
6. What indicators does an earth fault instrument give?

11 ELECTRICAL SAFETY

11.1 Questions – General Electrical Maintenance

1. What should alternator windings be checked for?
2. What should alternator terminal boxes be checked for?
3. What should alternator air-ducting be checked for?
4. What must be guarded against when cleaning windings with low-pressure air?
5. What should alternator slip rings be checked for?
6. What should be done if the winding's resistance is less than $0.5M\Omega$?
7. What should be done if an alternator is laid up for a long period?

11.2 Questions – Switchboard Maintenance

1. What precautions must be taken when carrying out work on switchgear?
2. Before starting work, what tests should be carried out?
3. What should the switchgear insulator be checked for?
4. How is the alignment of contacts checked?
5. What should be used to dress copper contacts?
6. What maintenance is usually carried out on silver plated contacts?
7. What can excessive application of petroleum jelly cause?
8. What are dash pots checked for?
9. How are trips and relays checked?

11.3 Questions – Motor Maintenance

1. What does dirt on motor insulation cause?
2. What happens when motor ventilating ducts are clogged?
3. In maintenance, what is next to godliness?
4. What can a layer of dust cause with totally enclosed motors?
5. Why must the air pressure be less than 1.75 bar when blowing out motors?
6. Name one source of oil and grease contamination in motors?
7. How should oil and grease contamination be removed from a motor?
8. What must be done when a motor is dismantled?
9. What could cause damaged insulation of the stator winding's insulation?
10. What could cause discolouration of the stator winding's insulation?
11. What could cause rubbing of a stator core?
12. What checks are made to a motor after changing bearings?
13. How is moisture detected in motor windings?
14. When should bearings be renewed?
15. What is the procedure if a bearing cannot be replaced?
16. What is the best way to clean and inspect motor bearings?
17. What must be done if a bearing sticks?
18. What visible signs would cause a bearing to be scrapped?
19. How much grease should be put in a bearing?

20. If the bearing housing has no vent hole, what must be done before regreasing?

21. Which application requires special greases for motor bearings?

22. What is a motor inspected for?

23. How is salt contamination of a motor removed?

24. How are stator windings usually dried out?

25. What tests are carried out before revarnishing?

11.4 Questions – Motor Control Gear Maintenance

1. What do all moving contacts have?

2. What could happen if 'wipe' is lost from contacts?

3. What condition do contacts need to be in before requiring filing?

4. What reduces mechanical wear?

5. What causes overheating with copper contacts?

6. How is rust removed from a magnet?

7. What should starter enclosures be checked for?

8. What should be checked in high vibration areas?

9. Should fixed and moving contacts be replaced separately?

11.5 Questions – Maintenance of Lighting

1. What must be ensured when replacing lamps or tubes?

2. What can arc flash cause personnel?

3. What can happen if a higher than rated light is fitted in a lamp fitting?

Engineering Pocket Book – Questions

    4.   What must be remembered when changing fluorescent tubes?

    5.   What should be checked with portable cargo lighting?

    6.   What will corrosion cause in flameproof light enclosures.

    7.   What should be done if the lamp glass is cracked in a flameproof enclosure?

    8.   What could happen if bolts are overtightened?

11.6 Questions – Electric Shock

    1.   What is the maximum current for 'let-go'?

    2.   What is the maximum voltage regarded as reasonably safe for portable power tools?

    3.   What is the maximum shock voltage to earth for a centre tapped 110V tool transformer?

    4.   What is the first thing that must be done when finding a victim of electric shock?

    5.   Give two methods of how this is done.

    6.   Where can details be found of the first aid to be carried out in the event of electric shock?

11.7 Questions – Circuit Safety Devices

    1.   What circuit protection devices are fitted on switchboards?

    2.   Under fault conditions, what should be the last protection device to operate?

    3.   What type of fuse should be used?

    4.   Give two methods of tripping non-vital loads to ensure essential supplies.

11.8 Questions – Earth Safety Devices

1. What stops an earth fault at a transformer?

2. Why must an earth return path have low impedance?

3. At what value of earth currents are earth leakage breakers set?

11.9 Questions – Welding

1. What is the range of welding current?

2. What are the two main non-electrical safety precautions used to protect welders?

3. What is used to protect operators from indirect electrical shock?

4. Why is the type B electrode holder better than the type 'A'?

11.10 Questions – Static Electricity and Interference

1. What is the classification of equipment that suffers from interference?

2. Give an example of non-electrical equipment giving rise to static.

3. Give an example of a very common source of interference which is a steady state condition.

4. What can cause intermittent interference?

5. What is the only guaranteed way of preventing interference?

6. How are radio receivers and transmitters constructed to reduce interference?

7. Where are suppressors fitted?

8. What types of cables should not be run together?

11.11 Questions – Emergency Power Supplies

1. Why must an emergency electrical power system be provided?

2. What must an emergency generator be provided with?

3. When must an emergency power source come on line?

4. Why are both an emergency generator and batteries recommended?

5. How long must battery powered lightening run on passenger ships?

6. When is the testing of emergency power supplies done?

7. What are the special interlocks fitted to the main and emergency generation circuit breakers for?

8. What are two methods of auto-starting an emergency generator?

9. How is the auto-start tested?

10. What is the only way to test the performance of the emergency power supply?

11. What equipment is usually fed by batteries?

12. What are the different types of battery cell?

13. What is the range of voltages of a lead acid cell?

14. What is battery capacity rated as?

15. During charging, which gas is given off?

16. Why must the two types of cell be kept separate?

17. What does the value of the electrolyte specific gravity indicate in lead acid batteries?

18. What arrangement is used in battery charging equipment?

19. What is the maximum allowable temperature of the electrolyte during charging?

20. What is the only indication of a fully charged alkaline cell?

21. What will happen if the cell plates are exposed to air?

Engineering Pocket Book – Questions

12 INSTRUMENTATION AND CONTROL

12.1 Questions – Control (General)

1. What are the two major factors that have influenced the development of marine automatic controls?

2. What are the 12 main parameters monitored at sea?

3. What does the UMS class notation mean?

4. What does the CCS class notation mean?

5. What are the five different control media?

6. What is the process of long distance control signal transmission called?

7. What is the modern definition of a transducer?

12.2 Questions – Control Systems

1. What is a control system?

2. How are control systems classified?

3. What provides information to the controller?

4. What is the term 'transmitter' commonly used to describe?

5. What does the behaviour and performance of a control system depend on?

6. What is a control system called if its action is dependent upon the output?

7. What is excessive corrective action called?

8. What is the difference between the input and output signals called?

9. What are the three common essential elements of a control system?

10. What are the three components of a transducer?

11. What is the set point value?

12.3 Questions – Measuring Instruments

1. What are the three methods of measuring pressure?

2. What are the six factors affecting the deflection of a Bourdon Tube gauge?

3. What are the five factors affecting the deflection of a diaphragm sensing element?

4. What is used to increase the spring rate of a diaphragm capsule stack?

5. What materials are diaphragm sensing elements made from?

6. What method is usually used at sea for level measurement?

7. Why is a second liquid sometimes used in a single column level indicator?

8. What must a diaphragm level indicator be mounted clear of?

9. When is a dip tube level indicator used?

10. What is measured to give the tank level with a dip tube level indicator?

11. How is a 'Pneumercator' Tank Gauge operated?

12. Why is a mercury trap fitted to the manometer of this gauge?

13. What must the bell of this gauge be clear of?

14. What are the minimum lengths of straight pipe before and after an orifice plate?

15. What can cause the upstream straight section to be increased?

16. What must the leading edge of an orifice plate be?

17. What type of pressure tappings are used for an orifice plate in a small-bore pipe?

18. When measuring fuel oil flow with an orifice plate, what is fitted between the differential pressure instrument and the pressure tappings?

19. What must be done before zeroing a control system transmitter?

20. How does a variable area flowmeter work?

21. What is the minimum flow that can be measured through a variable area flowmeter?

22. What must be carefully selected for a variable area flowmeter?

23. What is used to calibrate a turbine flowmeter?

24. What is possibly the most important variable monitored at sea?

25. What causes errors in temperature measuring systems?

26. How does a bi-metallic thermometer operate?

27. What is the most common form of Bourdon tube in a mercury-in-steel thermometer?

28. What is the most popular type of distant reading thermometer?

29. Why are helium and nitrogen popular for gas-filled thermometers?

30. What is the main disadvantage of a gas thermometer?

31. How is a resistance thermometer temperature element constructed?

Engineering Pocket Book – Questions

32. What can the temperature element be sealed in?

33. In a resistance thermometer, what is the bridge resistance made of?

34. What are used to eliminate the error caused by the ambient temperature in a two wire resistance thermometer?

35. Which is the most popular method?

36. What does a high bridge current do to a resistance thermometer?

37. Which type of bridge is the more accurate in a resistance thermometer?

38. What connects the cold junction of a thermocouple to the measuring instruments?

39. What is usually the minimum range of a moving coil millivoltmeter?

40. What is the most common method of eliminating the effect of ambient temperature from moving coil temperature indicators?

41. What are the three methods of connecting thermocouples?

42. How is the fastest possible response from a thermocouple obtained?

12.4 Questions – Automatic Control Theory

1. What is feedback?

2. What is meant by open loop control?

3. What is an example of two step control?

4. What may occur in two step control?

5. What is another name for modulating control?

6. What is the basic form of modulating control?

7. How is the best performance of a control system and plant obtained?

8. What is adjusted to maintain control system performance?

9. What is meant by a proportional band of 100%?

10. What is the gain of a controller?

11. With proportional control only, what happens when there is a change in plant loading?

12. What kind of error is offset?

13. How long will an offset remain?

14. What is the difference between the desired value and the set point called?

15. How often would the set point be adjusted to correct the droop of a proportional controller?

16. How is the offset reduced in proportional control?

17. What is oscillation of the correcting element called?

18. What is the name given to integral action?

19. What is another name for instability in a control system?

20. How is hunting avoided with a proportional controller?

21. How is the reset effect measured?

22. Why should the adjustment of integral action be gradual?

23. What is the name given to derivative action?

24. What is used to eliminate hunting when there are long reset times?

Engineering Pocket Book – Questions

25. What is derivative action time?

26. What is the integral action time when the integral action is large?

27. What type of control uses two or more correcting elements for the same controller output?

28. What is used to avoid problems with two open control valves when using split range control?

29. What type of control keeps two variables in fixed proportions?

30. What is cascade control?

31. What are three types of delay in system response?

32. How is distance velocity lag reduced?

33. What is transfer lag sometimes known as?

12.5 Questions – Types of Control Action

1. What condition of the system response is one of the most important design considerations?

2. Where do most failures in control systems occur?

3. What could cause a controlled condition to go to one extreme of its range?

4. What could cause a controlled condition to stay at zero?

5. What could cause a controlled condition to oscillate?

12.6 Questions – Automatic Controllers

1. For the force balance beam type controller illustrated, what is the comparing element?

2. What is the position of the beam pivot for 100% proportional band?

3. How is the gain increased?

4. What is applied to a controller to eliminate offset?

5. What does the integral action time depend on for the beam type controller illustrated?

6. What is used to give rate action to the beam type controller illustrated?

7. What is added to a controller to overcome time lag?

8. To increase the reset action, which way is the restrictor valve turned?

9. What type of controller is most common at sea?

10. For the nozzle/flapper controller illustrated, how does the input signal alter the output pressure?

11. For minimum feedback, where should the pivot be in the nozzle/flapper controller illustrated?

12. Which bellows give negative feedback?

13. Which bellows give positive feedback?

14. For a stacked type controller, what is usually the ratio of the areas of the diaphragms?

15. How is the response adjusted for the stacked controller illustrated to suit plant requirements?

16. What is a typical current range of an electronic controller?

17. What happens when an electronic controller's amplifier drifts off value?

18. Why are AC amplifiers preferred for electronic controllers?

19. How is the AC output signal used for control purposes?

20. How is reset action time varied on the electronic controller illustrated?

21. What is the difference between a pulse controller and the above types?

22. What does the pulse controller allow when changing from auto to hand?

23. What must be done before making adjustments to a controller?

24. On a proportional controller, how is the proportional band set?

25. What is optimum proportional band sometimes called?

26. How is a two term controller checked for offset?

27. Before settling, how many oscillations should the plant make with a controller or Critical Proportional Band?

28. In setting a two term controller, what is the next procedure after the periodic time has been found?

29. Before settling, how many oscillations should a plant make with a two term controller properly set?

30. When a three term controller is properly set, what should be the plant response to a change?

31. How does a pneumatic remote set point adjuster operate?

12.7 Questions – Control System Components

1. What is an error detector?

2. What are the two classes of error detector?

3. What are mechanical differentials used for in rotating systems?

4. What causes dead band in a mechanical differential error detector?

5. What are examples of control systems with positional gyroscopes?

6. Why does a positional gyroscope require periodic resetting?

7. What are examples of rated gyroscope measurements?

8. What are potentiometers used for in error detection?

9. What is used for error detection in temperature control systems?

10. What are the four classifications of controller?

11. What is the main disadvantage of mechanical error detectors and controllers?

12. What are the three types of electric amplifiers used in control systems?

13. How is motor speed regulated in a pump controlled hydraulic system?

14. What is an example of a controller in a valve-controller hydraulic system?

15. Where are low pressure pneumatic control systems used?

16. What is the essential difference between hydraulic and pneumatic control systems?

17. What type of pneumatic controller is used for displacement measurement?

18. What type of controller is found in high pressure pneumatic control systems?

19. In which control systems are DC electrical motors used as output elements?

20. What is the main advantage of field control of DC electrical motor output elements?

21. How much extra pressure may be required to operate a hydraulic piston actuator?

22. What are low pressure pneumatic activators known as?

23. What are low pressure pneumatic actuators used for in the process industry?

12.8 Questions – Supply Systems

1. What is meant by instrument air?

2. How small can a particle be to block orifices and nozzles?

3. What can moisture cause in a pneumatic system?

4. How is instrument air dried?

5. How is air dried in an absorption drier?

6. What reduces the absorption efficiency of a dessicant?

7. What is the purpose of the air dryer outlet filter?

8. How are the dryers changed over in the system illustrated?

9. What is involved in the process of regeneration?

10. Where does the auto-unloader's operating air usually come from?

11. What is the purpose of an auto-unloader?

12. How does a filter regulator operate?

13. Where does the filtering of power lines take place?

12.9 Questions – Practical Control Systems

1. How can an appreciable fuel economy be effected on turbine driven ships?

2. For the temperature control system illustrated, what is the transmitter's output signal range?

3. What is adjusted to restore the oil temperature to its desired value?

4. What is the difference between a temperature control system for lubricating oil and one for boiler fuel oil?

5. For the auxiliary exhaust steam control system illustrated, what is the steam pressure converted to by the pressure transmitter?

6. What prevents high pressure steam being dumped to the condenser with this system?

7. For the simple combustion system illustrated, what is measured by the pressure transmitter?

8. Where is the controller output taken?

9. What adjustment is made to the forced draught controller to obtain the desired combustion?

10. How is the differential pressure across the fuel control valve maintained?

11. What should be measured for accurate combustion control of high capacity boilers?

12. For the main boiler combustion control system illustrated, what type of controller receives the steam pressure signal?

13. What is used to combine the steam load when first detected?

14. Where is a rapid change in steam load first detected?

15. Where else does the signal to the burner fuel oil control valve go?

16. What is the source of the measured value to the air flow controller?

Engineering Pocket Book – Questions

17. What is the desired position of the forced draught vanes?

18. What happens in practice at the forced draught fan with an increasing load?

19. How is excess air ensured with a decreasing steam load?

20. Why does the drum water level rise with increased steam demand?

21. What type of boilers are single element feed control systems limited to?

22. Why is only steam flow measurement only insufficient to control the drum water level?

23. How is the steam flow measured?

24. What does the water level controller do to the master signal to the feed water control view?

25. For the diesel engine lubricating oil system illustrated, where does the controller output go?

26. What is the purpose of the solenoid valve in this system?

27. For the jacket water cooling system shown, what type of valve actuation is used?

12.10 Questions – Alarm Indication Systems

1. What is a basic alarm system?

2. How does a pressure or temperature operated electrical switch work?

3. What should be done after 'accepting' an alarm?

4. Name three types of alarm indication on an electrical alarm system?

5. What kind of alarm indicator is used with a pneumatic alarm system?

6. What are the two scanning speeds of a scanning alarm system?

7. What is displayed during the fast scan?

8. What is the basic measuring system for temperature?

9. What transducers are used for pressure measurement?

10. What is fitted between a contact alarm and the alarm annunciator?

11. What are the three basic functions performed by the scanner at each selected point?

12. How does the converter operate?

13. Why are the converter and the central processor isolated?

14. What are the sub-systems of the central processor?

15. What resets the circuitry for each measurement?

16. What is the timing circuit responsible for?

17. Where are the alarm levels 'set'?

18. What happens when no alarm is detected?

19. When are groups of annunciators switched out?

20. What happens when the self-check system finds a fault?

21. How often should the self-check system be tested?

12.11 Questions – Operation And Maintenance

1. Why should the equipment manufacturer's operating instructions be follows?

2. What kind of transfer should be obtained when switching from manual to auto?

3. How should control cabinets be designed for maintenance?

4. What must a differential pressure type flow measuring instrument have for accurate results?

5. Where should pressure tappings not be made?

6. What, in a control system, should a pressure tapping keep to a minimum?

7. Why are bonded strain gauge type transducers best for marine use?

8. Why are most control valves mounted vertically?

9. At what position is it an advantage to fit a diaphragm operated control valve?

10. What can happen if the signal lines are wrongly routed?

11. What can happened to electric cables in high temperature areas?

12. What type of piping should be used in machinery spaces for pneumatic control systems?

13. Why should pneumatic piping be run like electric cables?

14. What problems should be considered when laying out a control system?

15. What hazard is reduced by fitting receiver type gauges instead of direct reading gauges in control panels?

16. What should be done after the installation has been inspected for damage?

17. Why is a system cold run trial vital?

18. What instrument calibration accuracy can be obtained in manufacturer's tests?

19. What is absolutely essential for pneumatic systems?

20. What is the usual maintenance procedure with equipment of solid state modular construction?

21. How is poor performance recognised?

22. What must be done before fault finding?

23. In fault finding, what can make the situation worse?

24. What is vital in order to realise the full benefits of automation?

25. What can an increase in testing mean to a control system?

26. What length of time should a typical maintenance schedule be based on?

27. What does good documentation of regular testing do?

13 DIESEL ENGINE THEORY & STRUCTURE

13.1 Questions – Engine Types

1. What are the two main types of diesel engine?

2. Which type tolerates low quality fuel best?

3. Which type dominates the ocean-going propulsion market?

4. Up to what value is the bore/stroke ratio of long stroke, slow-speed engines?

5. Where are medium speed engines widely used?

6. What are the different types of turbocharge system?

7. Which type of turbocharge system gives the best overall efficiency?

8. On what principle did the original diesel engine operate?

9. On which theoretical cycle do modern diesels operate?

10. What are the three reasons why the actual work done diagram differs from the ideal work done diagram?

11. What does a 'draw card' give an approximation to?

12. What are the four operations within a diesel engine cylinder?

13. What is an engine's stroke?

14. What does engine timing refer to?

15. What does TDC mean?

16. What does BDC mean?

17. What does scavenging mean with regard to two-stroke engines?

13.2 Questions – Engine Construction

1. What are the main bearings of direct drive engines aligned with?
2. What material is used for chocks instead of cast iron or steel?
3. How are holding down bolts secured to the tank top?
4. Why is there clearance between the holding down bolts and the bedplate?
5. How are holding down bolts tightened?
6. How are chocks tested?
7. What supports the main bearings and crankshaft?
8. What does the casting in the centre of the transverse web support?
9. How can the overall height of an engine be limited?
10. What type of welds are used in bedplate construction?
11. Where do fatigue cracks start?
12. Why are the engine frames known as 'A' frames?
13. What passes up through the frame?
14. What is the main strength member of a medium-speed engine?
15. How is the engine structure pre-stressed?
16. Why should the tie bolt centres be as close as possible to the crankshaft axis?

13.3 Questions – The Crankshaft

1. Where do the main bearings support the crankshaft?
2. What properties must the crankshaft's material have?

3. What process improves fatigue resistance in medium-speed crankshafts?

4. How does oil get to the big end bearings in a medium-speed engine?

5. Describe how a semi-built crankshaft is constructed?

6. Why are dowels and keys not used in semi-built crankshafts?

7. What does overlap permit in a welded crankshaft?

8. What two methods are used to improve the balance of the crankshaft?

9. What kind of bearings are used in large crosshead engines?

10. Where is the thrust bearing usually located in direct drive engines?

11. What type of bearing is used as a thrust bearing?

12. On slow-speed engines, what is the flywheel for?

13. What must be done before turning the engine?

14. What does the turning gear safety cut-out ensure in the manoeuvring gear?

13.4 Questions – Connecting Rods and Crossheads

1. In what direction is the oil flow in the connecting rod of a crosshead engine?

2. Why is the connecting rod kept as short as possible?

3. What shape is the section of a medium-speed engine connecting rod?

4. What are engine guides made of?

5. How are the guides lubricated?

6. What does excess clearance between engine guides and shoe cause?

7. What material is used to form the crosshead?

8. What style of bearing is used as top end bearings?

9. With the older design of crosshead, where is the main concentration of load?

10. What is the equivalent of the crosshead in trunk piston engines?

11. Why may a non-return valve be fitted at the foot of a medium-speed engine's connecting rod?

12. What do large diameter crank pins improve?

13. What are the bearing materials used in medium-speed engine bottom end bearings?

14. What do the harder bearing materials require in a bearing?

15. Why are the angled butts of a medium-speed engine connecting rod serrated?

16. What are the three different arrangements of big end bearings on a crank pin in Vee engines?

13.5 Questions – Pistons

1. What other components require similar material properties to a piston?

2. What factors do the choice of material and design of pistons depend on?

3. What steel is used for pistons in two-stroke engines?

4. What is done to pistons to give extra protection from high temperature corrosion?

5. Which coolants are used for pistons in slow-speed engines?

6. Which piston coolant has the higher thermal capacity?

7. Why is the piston cooling system kept running after 'finish with engines'?

8. What is the main purpose of the skirt on a large piston?

9. What is fitted to skirts of large pistons to assist running in?

10. Why are long skirts fitted in some designs of slow-speed engines?

11. What material is used for the skirt in slow-speed engines?

12. What is the purpose of the piston rod?

13. Why is the piston rod round in section?

14. How is the piston rod attached to the piston?

15. Why may a piston rod be surface treated?

16. Why is the top surface of a four-stroke piston recessed?

17. What dictates the length of a four-stroke piston skirt?

18. Why are aluminium alloy pistons not used when burning residual fuels?

19. To what shapes are piston rings machined?

20. What are the three types of piston ring joints?

21. What are the two main purposes of piston rings?

22. What increases the outward pressure of the piston rings?

23. What should be done when wear of the piston rings becomes excessive?

24. How is the circumferential clearance of piston rings measured?

25. How is the circumferential wear of piston rings measured?

26. What coatings are applied to piston rings?

27. Which kind of cylinder liner is not used with chromium or plasma coated piston rings?

28. Why do some piston rings have barreled or grooved surfaces?

29. What is the other name for an oil control ring?

30. What is the purpose of oil control rings?

13.6 Questions – Cylinder Diaphragm and Piston Rod Gland

1. What contamination is prevented by a diaphragm?

2. What does a diaphragm mean with regard to engine lubrication?

3. What is the purpose of the upper half of the piston rod gland?

4. What is the purpose of the lower half of the piston rod gland?

5. What can the lack of or incorrect maintenance of the piston rod gland cause?

13.7 Questions – Cylinder Liner

1. What is the base material of the cylinder liner?

2. What are the alloying elements vanadium and titanium added for?

3. What are the two methods used to retain lubrication with chrome plated liners.

4. How is a cylinder liner secured?

5. Why can the thickness of a cylinder liner be reduced at the lower end?

Engineering Pocket Book – Questions

6. Why are the port edges shaped?

7. How is the lower part of the cooling water space sealed?

8. How is the upper part of the cooling water space sealed?

9. What does the term 'bore cooling' mean?

10. In large engines, what does the cylinder jacket consist of?

11. Why do cylinder liners require lubrication?

12. For a heavy fuel oil burning crosshead engine, what is the TBN of the cylinder oil?

13. What properties must a cylinder oil have in a crosshead engine?

14. Why must a cylinder oil not be reused in a crosshead engine?

15. What supplies the cylinder oil to the lubrication quill?

16. How is the blowback of the combustion gas prevented in a cylinder oil lubrication system?

17. How does an accumulator work in a cylinder oil system?

18. How is the cylinder oil distributed over the length of the cylinder in a crosshead engine?

19. When should the supply of cylinder oil be increased?

20. With medium-speed engines, how is cylinder lubrication usually achieved?

21. How is piston cooling oil used for cylinder lubrication in medium-speed engines?

22. In medium-speed engines, why is insufficient cylinder lubrication dangerous?

23. What will excess cylinder lubrication cause?

Engineering Pocket Book – Questions

13.8 Questions – Cylinder Cover

1. What is the other name for a cylinder cover?
2. What valves are usually found in a slow-speed engine cylinder cover?
3. Where does the cylinder cover cooling water come from?
4. How are medium-speed engine cylinder covers made?
5. What passages are usually found in medium-speed engine cylinder covers?

13.9 Questions – Camshafts and Valve Gear

1. What operates the valves and fuel pump of an engine?
2. What treatment is done to cams?
3. Why do some engines have a separate camshaft lubrication system?
4. What must be checked periodically with camshafts?
5. What do the cams on a four-stroke engine operate?
6. At what speed does a slow-speed engine camshaft operate?
7. What drives the camshaft?
8. What are the two types of camshaft drive?
9. Which camshaft drive is used when the distance between the crankshaft and camshaft is large?
10. Which camshaft drive gives a reduction in weight?
11. Why is there a high factor of safety in chains?
12. Where are the three areas that wear occurs in chain drives?

Engineering Pocket Book – Questions

13. How is wear in the chain measured?

14. What should the transverse movement be roughly equal to?

15. What is the approximate life of a camshaft chain?

16. Why does one jockey wheel have an adjustable centre?

17. What is usually the maximum adjustment of a chain?

18. How many times can a pin link be re-riveted?

19. How is the timing adjusted?

20. How is the timing checked?

21. What will excessive tension cause in a chain drive?

22. What types of valves are used for exhaust valves?

23. In four-stroke engines, how are the valves actuated?

24. What material is used to make exhaust valves?

25. What is stellite?

26. What is used to protect the underside of some large exhaust valves?

27. How are exhaust valves cooled?

28. What does the rotation of valves ensure?

29. Why are tappet clearances necessary?

30. What will tend to increase the tappet clearance?

31. What will excess clearance cause?

32. What can insufficient clearance cause?

33. What will be caused if the combustion gases are allowed to blow past the valve seat?

34. What does the hydraulic system replace in mechanically actuated valves?

35. How is the pressure in an air spring maintained?

36. What size should air inlet valves be?

13.10 Questions – Turbocharge Systems

1. Why are engines turbocharged?

2. How does the charge air cooler improve the engine thermal efficiency?

3. When is a turbocharger said to be 'matched' to an engine?

4. What are the two operating systems for turbochargers?

5. Which system gives a rapid build-up of turbine speed when starting or manoeuvring?

6. Which system gives high thermal efficiency?

7. What does a divided exhaust system do to the system efficiency?

8. What are the two parts of a turbocharger?

9. How are the turbine blades attached to the turbine disc?

10. What material is used for the turbine blades and nozzles?

11. What material is used for the air impeller and inducer?

12. What is fitted at the air inlet?

13. Which four parts are changed together to ensure matching?

14. What type of seals are fitted to the shaft?

15. What do the seals prevent?

16. At which end of the shaft is the thrust bearing fitted?

17. What is the usual life of a roller bearing?

18. What causes a turbocharger to rotate when the engine is out of service?

19. What is cooled in an 'uncooled' turbocharger?

20. How does an uncooled turbocharger increase the thermal efficiency of a ship?

21. Why are lower compression temperatures preferred?

22. How is the air velocity increased after the charge air cooler?

23. What can be indicated at the charge air condensate drain?

24. How is the air seal maintained at the free end of the charge air cooler?

25. What should be the maximum air discharge temperature from the charge air cooler?

26. What can too low an air temperature cause?

27. What does two-stage turbocharging give?

28. Above what temperature should charge air be kept to prevent undercooling?

14 FUEL SYSTEMS

14.1 Questions – Fuel Oil

1. What type of fuel is used in modern marine diesel engines?

2. What are the three major problems associated with modern marine fuels?

3. What is the reason for the increase in sludge building up in modern marine fuels?

4. What kind of additive is used to reduce the formation of sludge?

5. What can water in the fuel system do?

6. How is water removed from bunker tanks?

7. What is the significant problem caused by poor or incomplete combustion?

8. Give five problems arising from burning heavier fuels?

9. What do these problems cause in a diesel engine?

10. Give four qualities used to indicate a fuel's burnability?

11. What is important in order to achieve atomisation particularly at low load?

12. What is the major fuel content influencing high temperatures corrosion?

13. What is the minimum melting temperatures of vanadium compounds of combustion?

14. Describe the action of a severe form of high temperature corrosion?

Engineering Pocket Book – Questions

15. What is the main defence against high temperature corrosion?

16. How is the prevention of high temperature corrosion achieved?

17. Describe the action of low temperature corrosion?

18. What is the danger of raising temperature to combat low temperature corrosion?

19. What are the normal abrasive impurities found in fuels?

20. What is the new type of contaminant being found in modern marine fuels?

21. What is the source of this contaminant?

22. What was once considered the best guide to fuel quality?

23. What is the viscosity of a fuel?

24. What is necessary in the handling of modern marine fuels and why?

25. When giving a viscosity, what must be stated?

26. What does a high cetane number indicate in a fuel?

27. What is another name for the Conradson carbon value?

28. What is the ash content of a fuel the measure of?

29. Why are high sulphur levels in a fuel dangerous?

30. How is the water content of a fuel determined?

31. What is the pour point of a fuel?

32. What is the flash point of a fuel?

33. What is the specific gravity of a fuel used to calculate?

Engineering Pocket Book – Questions

14.2 Questions – Supply of Fuel Oil

1. Why must the density of a fuel be given when bunkering?
2. What is the viscosity of the fuel used to calculate?
3. What happens in the setting tanks?
4. What is the recommended treatment of fuels?
5. What is the maximum temperature of the fuel during treatment?
6. To avoid fouling of the fuel line, what temperature should not be exceeded?
7. What mesh is the fine filter in the fuel line made of?
8. What must be included in the fuel oil system?
9. What is the maximum fuel storage temperature?
10. What must be reduced to achieve correct atomisation of fuel?
11. What is the viscosity of a fuel for combustion?
12. What is the general indication of good combustion?
13. What are the four parameters for combustion?
14. What does the fuel droplet size depend on?
15. What does penetration refer to regarding combustion?
16. What are the three conditions that penetration is dependant on?
17. What determines the spray pattern of the fuel?
18. What imparts swirl to the air in a cylinder?
19. Why is turbulence important in combustion?
20. What is the term used to describe the combustion in diesel engines?

21. What happens during the first phase of combustion?

22. What is the time taken during this first phase called?

23. What happens during the second phase of combustion?

24. What is diesel knock?

25. What does ignition quality denote?

26. What is the usual measurement of ignition quality?

27. Down to what cetane number can slow-speed engines operate?

28. What is the lowest cetane number for a medium-speed engine?

29. When is the ignition quality important?

30. At low speeds, what can the increase in ignition lag cause?

31. What does variable ignition timing do?

32. The design of what builds variable ignition timing into medium-speed engines?

33. What are the different methods of pilot ignition?

34. What is the most notable feature of an engine with electronic ignition?

14.3 Questions – Fuel Pumps

1. What operates the fuel pump plunger?

2. When does fuel delivery start on a jerk type pump?

3. When does delivery close?

4. What regulates the quantity of fuel in a jerk type pump?

5. How is injection timing controlled?

6. What are the two methods of varying the point of injectors?

7. How is the plunger lubricated?

8. What controls the delivery in a valve timed pump?

9. How is timing altered with a valve timed pump?

10. How is variable ignition timing achieved with a valve timed pump?

14.4 Questions – Fuel Injectors

1. What are the two basic parts of a fuel injector?

2. What joins these two parts?

3. When does the needle valve open?

4. What is necessary when the injectors are operating on heavy oil?

5. What ensures the alignment of the oil passages on injectors?

6. What should be done with fuel injectors prior to sailing?

7. What defects are found in injectors?

8. What must be done with exposed fuel pipes?

9. What happens if the injection tip temperature is too high?

10. What is used for cooling in modern injectors?

11. What is the ideal position for the fuel injector?

12. What can happen when running at low power for long periods?

15 COMBUSTION

15.1 Questions – Fuel and Fuel Burning (Diesel Engines)

1. Explain the importance of viscosity.

2. Explain the importance of flash point.

3. Explain the importance of ignition point.

4. State the viscosity for diesel oil and heavy residual fuel.

5. What is the importance of the cetane number and how is it measured?

6. State the problems caused by high sulphur content in fuel oil.

7. State the problems caused by high vanadium content in fuel oils for diesel engines.

8. Give reasons why vanadium cannot be removed from fuel oils for diesel engines.

9. Sketch a draw card and power card for a slow-speed engine cylinder with a leaking fuel injector.

10. What are the consequences of leaking fuel injectors?

11. Sketch the power card and draw card for after-burning in a diesel engine.

12. State how after-burning might occur.

13. How is power balancing accomplished on medium-speed diesel engines?

14. How are compression diagrams taken on slow-speed diesel engines?

15. What are the consequences of an engine operating in an unbalanced condition?

16 LUBRICATION

16.1 Questions – The lubricating oil system

1. Which items are duplicated in a lubricating oil system?
2. Where does the lubricating oil pressure pump draw from?
3. Where does the pump discharge to?
4. Where is the oil distributed to after the filters?
5. What are the drain returns kept clear of?
6. Why are the drain returns submerged?
7. Where is the drain tank generally situated?
8. Why is there a cofferdam around the drain tank?
9. Where is the level gauge situated to reduce reading fluctuations?
10. Why are the drain tank interior surfaces coated?
11. What is installed to improve the condition of the drain tank contents?
12. What oil contaminant must be eliminated or kept to a minimum?
13. Roughly how much extra oil is used for cooling slow-speed pistons?
14. What is the range of pressure for slow-speed engine lubricating systems?

16.2 Questions – Lubricating Oils (1)

1. How are the base stocks of lubricating oils obtained?
2. What are compound lubricating oils?

3. Why are compound oils used in the presence of water and steam?

4. What property do fatty oils gain when sulphurised?

5. What can be caused in feed systems and boilers due to fatty acid lubrication?

6. What is the total base number?

7. Why must the acidity of an oil be monitored?

8. What does demulsibility refer to?

9. What does corrosion inhibition relate to?

10. What is the condition known as scuffing?

11. What is often added to maintain the oil film under very severe load conditions?

12. What lubricants are used to prevent scuffing?

13. What is emulsification associated with?

14. What does oxidation do with regard to emulsions?

15. What leads to deposits?

16. On cool surfaces, what do deposits form?

17. How do additives increase the life of a lubricating oil?

18. A doubling of the oxidation rate is caused by how many degrees rise in temperature?

19. Oxidation of oil causes what change in the oil?

20. What does foam in the oil do to lubrication?

21. What is a corrosion inhibitor?

22. What does a detergent in an oil do?

23. What are dispersants in an oil?

24. When is a dispersant more effective than a detergent?

25. How does a pour point depressant work?

26. What conditions in the system can lead to the oil foaming?

27. What does a viscosity index improver do?

28. What is the action of an oiliness and extreme pressure additive?

16.3 Questions – Lubrication Fundamentals

1. What are the functions of a lubricant?

2. What happens to the friction force with a slight trace of lubricant?

3. What are the two most important properties of a lubricant?

4. What can be added to improve the oiliness of an oil film?

5. What determines friction when there is no material contact?

6. Where in the crankcase is boundary lubrication said to exist?

7. What is hydrodynamic lubrication?

8. Where is the maximum oil pressure in a bearing?

9. What are the four factors affecting hydrodynamic lubrication?

10. Why does insufficient circulation of lubricating oil mean high temperature?

11. If rotation speed is kept constant, how can relative speed be increased?

12. What can high bearing pressure lead to?

13. What kind of lubrication is found in a bearing as movement starts.

14. What happens to the oil in a journal bearing as the shaft speeds up?

15. In a Michell bearing, what does the bearing surface consist of?

16. What pressure order can be achieved in a Michell bearing?

16.4 Questions – Lubricating Oils (2)

1. What range of TBN have slow-speed engine crankcase oils?

2. What is done to measure the condition of a lubricating oil?

3. What does a blotter test indicate?

4. What has increased the problem of cylinder lubrication?

5. Why can local surface temperature be higher than the gas temperature at TDC?

6. What range of surface temperature is found at the top of the piston ring area?

7. What range of temperature is found at the lower end of the cylinder jacket?

8. Around what temperature does thermal cracking begin?

9. What are the six requirements of a cylinder lubricant?

10. What is the disadvantage of paraffinic oils as cylinder lubricants?

11. What wear rates are considered normal for cylinder liners for slow-speed engines?

12. What was the first type of cylinder lubricant developed for residual fuel burning engines?

13. What is the disadvantage of dispersion type cylinder lubricants?

14. What is the neutralising agent used in modern single phase cylinder lubricants?

15. What is the alkalinity of an oil expressed as?

16. Why is the speed of the neutralising reaction important?

17. What is meant by the spreadability of an oil?

18. Why is the storage stability of an oil important with regard to oil lines?

19. What percentage of the fuel consumption is the average cylinder oil consumption?

20. Why is chromium plating beneficial to wear rates?

21. Why is the oil flow rate to bearings greater than that needed for lubrication only?

22. Up to what percentage of additives can there be in a cylinder oil?

23. What is the contact pressure in modern bearings?

24. What is the drawback with the increasing viscosity of a lubricating oil?

25. What must be efficient to achieve low undercrown temperatures of pistons?

26. In piston oil cooling, what two conditions must be at a maximum?

27. What type of cooling does the motion of a piston generate?

28. What is formed with the oxidation of an organic mineral oil?

29. What is the result of thermal cracking of an oil?

30. How is thermal cracking prevented?

Engineering Pocket Book – Questions

31. What is cold sludge in a lubricating oil?

32. Why is sludge more likely to be found in trunk piston engine oil than crosshead engine oil?

33. At what temperature is lubricating oil centrifuging carried out?

34. What corrosion inhibitor does not work with salt water contaminated oil?

35. Why must oil additives not be water soluble?

36. Why must stable emulsions be avoided?

37. What viscosity index should a good crankcase oil have?

38. What is added to an oil to improve the load carrying capacity?

39. Why is the same oil used for cylinder and crankcase lubrication in trunk piston engines?

40. What does a poor quality fuel mean in the operation of an engine?

41. What determines the lubricating properties of an oil?

42. As the TBN increases, what content in the oil increases as well as alkalinity?

43. Which load carrying additives combat valve train wear and varnish formation?

44. What is the TBN of an oil used in high sulphur content fuel burning medium-speed engines?

45. What properties are essential for oil used in high-speed engines?

46. What oils are sometime preferred for use in lifeboat engines and emergency generators?

16.5 Questions – Shipboard Lubricating Oil Tests

1. Where is the only place a complete and accurate picture of the oil condition can be obtained?
2. Describe the procedure of the alkalinity test.
3. What gives the indication in the alkalinity test?
4. What is used to determine the oil condition from an oil spot on blotting paper?
5. What is indicated if there are contaminants at the centre of an oil spot?
6. What is the procedure of the Mobil Flowstick test?
7. What could cause a reduction in the viscosity of an oil?
8. What could cause an increase in the oil viscosity?
9. How are large amounts of water in oil detected?
10. What is formed in the oil when microbial degradation occurs?
11. What are oils with microbial infection prone to?
12. What is the maximum recommended water content for a detergent oil?
13. What should be done to jacket and piston cooling water as a precaution against microbial degradation of an oil?
14. What is added to oil to combat microbial infection?

16.6 Questions – Grease

1. What is a grease?
2. What fillers are used in grease?
3. What are the advantages of a grease?

4. What are sodium soap greases suitable for?

5. Which type of grease will emulsify?

6. What is the melting point of sodium soap grease?

16.7 Questions – Oil Drain Tank

1. Where is the oil drain tank situated in large engines?

2. What is the recommended width of the cofferdam around the oil drain tank?

3. What is the purpose of the cofferdam around the oil drain tank?

4. Why should the oil drain tank be as deep as possible?

5. How can the oil drain tank be made deeper than the double bottom?

6. Why is the bottom of the oil drain tank sloped and 'v' shaped?

7. Why is an inverted 'top hat' fitted below the lowest point of the oil drain tank?

8. What should oil returns be?

9. Enroute to the main return line, what does the piston cooling oil pass through?

10. Where in the tank are the oil returns situated?

11. Where is the oil pump suction in the oil drain tank?

12. What is prevented if oil is pumped out soon after returning to the oil drain tank?

13. How high above the oil drain tank bottom should the oil pump suction be?

14. Why should internal baffle plates be fitted in the oil drain tank?

15. Why are drain holes left at the bottom of the internal baffles?

16. How much ullage should there be in the oil drain tank with a full change of oil?

17. Why should strengthening ribs be fitted to the outside of the oil drain tank?

16.8 Questions – Pre-Cleaning and Corrosion Protection

1. What should be done to bulk oil tanks before putting them into the system?

2. What is essential when using protective paints?

3. What happens if a layer of moisture or rust forms before applying paint?

4. When is it better not to use protective paint?

5. In smaller engines with wet sumps, what is usually fitted into the engine lubrication system?

6. What is the purpose of the hand pump in the lubricating system of smaller engines?

7. What is the importance of cleaning the entire lubrication system?

8. Which cleaning agent must be removed after use?

9. How should grease and oil type preservatives be removed?

10. How long should flushing of the system be carried out for?

11. What is the normal temperature of the oil drain tank?

12. What is the most convenient method of heating the flushing oil?

13. Why is the heating of flushing oil most important with cast iron components of the lubricating systems?

14. When can the circulating of flushing oil be stopped?

15. After removing the temporary filters, what is the next procedure in the flushing of the lubrication system?

16. What kind of oil is used for flushing instead of a special oil?

17. How is the viscosity of an oil lowered?

18. Why should the same oil not be used for several flushing operations?

19. What is the increasing problem with lubricating oils onboard ship?

20. Why is a multiple tank not suitable for oil storage?

21. Why is a separate filling line essential for each oil storage tank?

22. Why are deck level flush filling lines with screwed plugs not recommended?

16.9 Questions – Rotary Displacement Pumps

1. Why do rotary displacement pumps have a lower efficiency than reciprocating pumps?

2. Why are rotary displacement pumps almost exclusively used for pumping oils?

3. What are the two types of rotary displacement pumps?

4. What do the number of threads in a counterscrew pump determine?

16.10 Questions – Coolers

1. What is the usual cooling medium used in coolers?

2. What is the oil in contact with in a tube cooler?

3. What are tubes usually made from in a boiler?

Engineering Pocket Book – Questions

4. What must be fitted in the water boxes of coolers if they are not made of steel or coated?

5. What may cause early tube failures in coolers?

6. How are the tube plates secured in a cooler at each end?

7. What is used to detect joint leakage without mixing the fluids?

8. How is the cooler shell manufactured?

9. Why is the shell material not critical?

10. Where should the seawater enter in a horizontal cooler?

11. Why are vent cocks fitted to coolers?

12. What is the recommended method of controlling the temperature of the oil in a cooler?

13. What method of temperature control gives a greater corrosion risk?

14. What can partial blockage cause in a tube cooler?

15. What kind of brush is used in the cleaning of cooler tubes?

16. When is chemical cleaning recommended in a tube cooler?

17. What should be done before handling chemicals?

18. What is required after using a chemical agent?

19. Where were plate type heat exchangers first developed?

20. What effect on heat transfer does the corrugation on the plates of a cooler have?

21. What acts as a barrier to heat transfer with smooth flow?

22. Why are tube cooler materials not suitable for plate coolers?

23. What is the usual jointing material in plate coolers?

24. How is the joint between the plates maintained?

25. What is used for high temperature joints?

26. What can over tightening of a plate cooler cause?

27. How are the flow ports arranged in a cooler plate?

28. What is used in the double-jointed section of a plate cooler to indicate leakage?

29. How much extra space is needed for dismantling a plate cooler?

30. How is the capacity of a plate cooler altered?

31. In which type of cooler is leak detection more difficult?

32. Why are lubricating oil coolers usually of the tube type?

33. What is the major drawback of plate coolers?

34. How are cracks detected in the plates of a plate cooler?

35. What are the main benefits of using titanium?

16.11 Questions – Oil Maintenance

1. What are the different types of contaminants in lubricating oil?

2. What contaminants are formed from the incomplete combustion of fuel and lubricating oil?

3. What are the five classes of filtration equipment?

4. For a given throughput, what is the relationship between filter size and filter mesh size?

5. What particle size is the largest passed by a Vokes multi-element filter?

Engineering Pocket Book – Questions

6. What are the elements in a Vokes multi-element filter made from?

7. Where are coarse lubricating oil filters fitted in the system?

8. Where are fine lubricating oil filters fitted in the system?

9. Until recently, what could the majority of fine lubricating oil filters not tolerate?

10. Up to what percentage of water contamination can an efficient purifier continuously handle?

11. Is the maximum rated throughput of a purifier the same as the optimum efficiency throughput?

12. Why are filters mounted in pairs?

13. How is a filter strainer cleaned?

14. What is usually the smallest particle size that can be removed by a wire mesh filter element?

15. What size of particles can get through an Auto-Klean filter?

16. Why are pressure gauges fitted before and after filters?

17. What filtering mediums are found in fine filters?

18. What kind of filter can be fitted in bypass lubrication lines?

19. What is the basic type of marine oil centrifuge?

20. What force acts on the particles in a centrifuge?

21. How do dirt particles move down the underside of the discs in a centrifuge against the oil flow?

22. What factors affect the size of particle removed by a centrifuge?

23. Up to what diameter are centrifuge bowls made?

24. How is the continuous operation of a centrifuge maintained over a long period of time?

25. What should be noted when overhauling purifiers?

26. What is the best setup for the continuous bypass purification of lubricating oil?

27. After engine shutdown, how long should the purifier be kept on line?

28. In a multi-engine installation, what is the minimum recommended number of lubricating oil purifiers?

29. What combines with lubricating oil water contamination to form sulphuric acid?

30. Why is pre-washing lubricating oil before centrifuging beneficial?

31. At what temperature should the washing water be?

32. Why is fine filtration still necessary with good detergent oils?

33. What can be brought into the engine with the induction air?

34. What happens as the solid content increases in a detergent oil?

35. What is a typical time between oil changes for medium/high and high-speed engines?

17 STARTING & MANOEUVRING SAFETY

17.1 Questions – Starting Air

1. What is the method used to start main propulsion engines?

2. What dictates the opening of an air start valve?

3. In modern practice, when is air introduced to the cylinder?

4. In indicator diagrams, what does the area under the curve represent?

5. Where should the non-return valve be fitted in the air start system?

6. What should be fitted between the non-return valve and the cylinder?

7. At what pressure is starting air stored?

8. What are the two connections on the air distributor for?

9. What keeps timing valves clear of the starting air cam?

10. In multi-cylinder Vee engines, what is a possible arrangement of starting the air valves?

11. How is a leaking air start valve detected when the engine is running?

12. What is the capacity (in starts) of starting air that must be stored for a reversible engine?

13. What is the preferred method of fitting the air receiver's fittings?

14. What must the pressure gauge on the receiver be directly connected to?

15. What size of drain valve should be fitted to an air receive?

Engineering Pocket Book – Questions

16. In what condition should air be received from the compressor?

17. What must be done before opening a receiver's manhole door?

18. What must be avoided during the internal clearing of an air receiver?

19. What must be looked for during internal inspection of an air receiver?

20. What precautions must be carried out when applying protective coatings?

21. How many starting air compressors must there be?

22. What is the maximum compressor delivery air temperature?

23. Why is manual starting attractive?

24. What type of drive is used in gear drive starters?

25. What voltages are electrical starters?

26. How are batteries connected for starting?

27. What are the two types of batteries used for electric starting?

17.2 Questions – Manoeuvring/Direct Reversing Engines

1. What must be reversed to manoeuvre a ship?

2. What timings must be adjusted on a four-stroke engine to run in reverse?

3. What are used on four-stroke engines to allow ahead and astern running?

4. On large two-stroke engines, what is adjusted to allow astern running?

5. Why are gearboxes fitted to medium-speed engine drives?

Engineering Pocket Book – Questions

6. What is a clutch?

7. With a unidirectional propulsion unit how many clutches are fitted?

8. What is the ratio of modern gearbox installations?

9. What type of gearbox is used in medium-speed drives?

10. What is the usual design for an internal gearbox clutch?

11. How are pinions attached to their shafts?

12. What kind of bearings were found in early gearboxes?

13. How is a controllable pitch propeller attached to the tailshaft?

14. How does the oil reach the operating mechanism of the controllable pitch propeller?

15. What was the duty of the early forms of 'inertia' type governors?

16. What is the fluctuation of the governor called?

17. Why are centrifugal governors not suitable for alternator driving engines?

18. What is the optimum speed of a centrifugal governor?

19. Why is there a tendency for wear to be over only a small area of the drive shaft?

20. What would happen if the governor spring were to fail?

21. What shape is a governor spring?

22. What is meant by droop?

23. What is the difference between fine and coarse droop?

24. What is used to indicate the load limit of an engine?

Engineering Pocket Book – Questions

25. Where should the governor be situated?

26. What happens when the oil in a governor is too thick?

27. What are the two factors essential for the production of generated voltage?

28. What is the usual droop found with marine governors?

29. What is used to provide droop in an AVR?

30. Why are overspeed trips fitted?

17.3 Questions – Safety Systems (Crankcase)

1. What governs the size of a crankcase explosion?

2. What is the normal operating atmosphere in a crankcase?

3. What is the most common cause of lowered lubricating oil flashpoint?

4. What is considered the minimum temperature for a 'hot spot'?

5. What range of air/oil mist ratio is most dangerous?

6. Is the primary or secondary explosion the most dangerous?

7. Why should there be no cross connections between crankcases?

8. What indicates a hot spot in a crankcase?

9. What is the best course of action when a hot spot is suspected?

10. What should be increased when a hot spot is suspected?

11. How long should a crankcase be allowed to cool before entering?

12. What should ensure that crankcase explosions never occur?

13. What is the principle behind oil mist detection?

14. Why are the oil mist sampling pipes inclined?

15. How often should the oil mist detector be tested?

16. What are the two duties of the crankcase relief valve cover?

17. What is the free area of the gauze fire trap?

18. What is the minimum combined area of the crankcase relief valve?

19. What are the three sides of the fire triangle?

20. What can cause ignition of a scavenge fire?

21. What other danger can a scavenge fire cause?

22. Why must the turning gear be engaged if the engine is stopped because of a scavenge fire?

23. What types of fire extinguishers are used in fighting a scavenge fire?

24. What could happen if the scavenge trunk is opened up too early?

25. What can be done to stop the natural flow of air through the engine?

26. What must be inspected after a scavenge fire?

27. What are two possible causes of starting air system explosions.

28. What is fitted to minimise the effects of an air system explosion?

29. What should be done if a leaking air start valve is suspected?

30. What should be done when the air system is not in use?

Engineering Pocket Book – Questions

31. What causes overheating of compressor discharge air?

32. At what pressure should a cylinder safety valve lift?

33. Why is the engine turned with indicator cocks open before starting?

34. What three alarms are crucial for engine operation?

35. What are the two sub-systems of an alarm monitoring system?

36. What happens if there is a failure of a sensor or broken cable?

37. Do all alarms have the same indicators?

38. Why must the recording device be of high speed?

39. What is a safety system?

40. Give seven parameters which will cause power to be reduced on an engine.

41. What type of system is known as a second stage protection device?

42. What dangers are caused by bilge water?

43. Where is the fire detection indicator fitted?

44. What should a control system be designed to do on failure?

18 BOILERS

18.1 Questions – Boiler Types

1. What are the two basic types of boiler?

2. Which boiler type is used for high-pressure, high-temperature, high-capacity applications?

3. What does main boiler design depend on?

4. What features are embodied in modern propulsion duty boilers?

5. Where were the superheaters fitted in early D-type boilers?

6. How many superheaters are usually fitted in modern boilers?

7. What is meant by reheat?

8. Describe a double evaporation boiler system.

9. What are the other names for a firetube boiler?

10. How are most firetube boilers now supplied?

11. What is the steam temperature at 7 bar?

12. What is the temperature difference across a waste heat unit?

13. Why is the gas outlet temperature kept about 180°C in an exhaust gas heat exchanger?

14. How much of the exhaust gas energy is used in waste heat recovery?

18.2 Questions – Evaporators

1. What is the most economically viable method of making fresh water from seawater?

Engineering Pocket Book – Questions

2. What is left in the heat exchanger after the vapour is taken off?

3. What happens when scale forms on heat exchanger tubes?

4. What should the maximum density of brine be?

5. How are scale deposits removed?

6. What cannot be used in the making of potable water?

7. What protects the evaporator shell from corrosion?

8. What materials are used in the heat exchanger section?

9. What materials are used for the demister mesh?

10. What happens if the distillate is above its required density?

11. What is the ratio of feed water to distillate?

12. What is done to prevent scale forming in the brine pipelines?

13. What further treatment does potable water require?

14. What is the maximum time between cleanings of an evaporator?

15. What temperature is recommended for feed water in a flash evaporator?

16. What is the gain ratio of an evaporator?

17. What type of multiple effect evaporation is usual for marine use?

18. What type of evaporation gives gain ratios of 8 or 9?

19. What type of mechanical compressor is preferred in vapour compression evaporation?

20. What must be done to the feed of a vapour compression evaporator?

18.3 Questions – Boiler Mountings

1. How many safety valves are required on a boiler?

2. Where are the safety valves fitted on a boiler?

3. How high must a valve lift for full flow?

4. What is the maximum allowable accumulation of pressure?

5. How is a conventional the safety valve adjusted?

6. What is the purpose of the safety valve easing gear?

7. What is feathering?

8. What is the full bore safety valve designed to avoid?

9. Describe the operation of a full bore safety valve.

10. What is the increase in throughput with a full bore valve compared an ordinary valve?

11. How do full bore safety valves solve high temperature problems?

12. How does an improved high lift safety valve operate?

13. How does a high lift safety valve reduce safety valve blow down?

14. What can water in the waste steam pipe cause?

15. What kind of valve is used for a boiler stop valve?

16. What is the difference between the main and auxiliary stop valve?

17. What is the purpose of the boiler stop valve?

18. Why are non-return valves fitted as feed check valves?

19. What is the location of the main feed check valve?

20. What indication must a feed check valve give?

21. What can happen if the boiler water level is too high?

22. Why do Classification Societies demand feed water regulators on watertube boilers?

23. What is the usual arrangement for water level indicators?

24. What action must be taken if the level disappears out of the gauge glass?

25. Why must the fuel oil cut off be manually reset?

26. What prevents deposits forming around the needle valve of the low level protection device?

27. What is the purpose of the blow down valve?

28. Why are scum valves fitted to boilers?

29. What is the purpose of an air vent?

30. When is the superheater circulating valve closed?

31. Why are water coolers fitted after the salinometer valve?

32. What is the most common type of on-load cleaning?

33. What has to be done to clean tubes if on-load cleaning is not done?

34. Why can using steam for on-load cleaning be expensive?

35. What is the simplest method of steam drying fitted in a boiler drum?

18.4 Questions – Water Gauges

1. What is the limit for using test cocks as a water level indicator?

2. How is the tubular gauge glass held and sealed?

Engineering Pocket Book – Questions

3. What is used to prevent water leakage if the gauge glass breaks?

4. Why are plate glass guards fitted to tubular gauge glasses?

5. Why is a diagonally striped board placed behind a gauge glass?

6. Which side of the gauge glass is blown through first?

7. What causes the water level to rise if the water cock is blocked?

8. What must be ensured when refitting the cocks in a gauge glass?

9. What can happen if the tubular glass is too short?

10. What can cause a build-up of deposits in a gauge glass cock?

11. Why must additional care be taken if the gauge glass is indirectly mounted when 'blowing down'?

12. After 'blowing down', which side of the gauge glass is opened first?

13. What should be done if the water level falls to the bottom of the gauge glass?

14. When must a gauge glass always be tested?

15. How does a reflex gauge glass work?

16. Why are reflex gauge glasses operated by chains or rods?

17. At high pressure, what does hot water do to glass?

18. What is used to protect glass at pressures above 34 bar?

19. What must be done when overhauling a high pressure gauge glass?

20. Why are the slots in the louver plate angled upwards in high pressure gauge glasses?

21. Why must the top and bottom of a double plate gauge glass be marked?

Engineering Pocket Book – Questions

22. If remote water level indicators are fitted on the same connections as a direct-reading gauge glass, what must be done before blowing the gauge glass?

18.5 Questions – Boiler Operation

1. What are the four extremely hazardous conditions to which all boilers are subject?

2. What is the prime consideration in boiler operation?

3. What can be the major heat loss in boiler operations?

4. How can the atmospheric heat loss be kept low?

5. What must be closely controlled for maximum efficiency and perfect combustion?

6. What is the main source of heat loss when there is insufficient air for complete combustion?

7. What funnel exhaust state is considered good practice?

8. How are unburnt losses in a boiler kept to a small value?

9. What happens to fuel oil in a settling tank?

10. Why are the burner connections from the fuel main kept as short as possible?

11. Why must there be no dead legs in the fuel main?

12. What is an essential safeguard for combustion equipment?

13. What should be done when wear is detected in a burner tip?

14. What should the boiler air trunking be checked for?

15. What increases the tendency for fouling in a boiler?

Engineering Pocket Book – Questions

16. What must be kept in working order to prevent problems with a boiler?

17. What is checked for corrosion from combustion products?

18. What is done to ensure gas side cleanliness during a shutdown?

19. Where is gas side cleanliness most essential?

20. Why are water treatment chemicals added?

21. What kind of cleaning is currently used on the waterside of a boiler?

22. What follows all boiler water side cleaning operations?

23. What is the purity of feed water related to?

24. What is customary practice after firing stops to improve circulation?

25. What should be checked for internally before filling the boiler?

26. What should the gas side of a boiler be checked for after maintenance?

27. What is done before closing the manhole?

28. What should be done before firing the boiler?

29. What is the final check before firing a boiler?

30. Why is care needed in the initial period of firing?

31. When is the superheat circulating valve closed?

32. What should be done when steam pressure reaches about 10 bar?

33. What is the time scale to go from flashing up to going on line?

34. When should this procedure be carried out more slowly?

35. How are modern boilers controlled?

36. Why must the firing rate be limited when regaining pressure?

37. What can happen if deposits accumulate on the heat recovery surfaces?

38. What can cause an accumulation of soot?

39. What factors influence the likelihood of superheater tubes overheating?

40. How is the free hydrogen formed for a hydrogen fire?

41. How is a hydrogen fire extinguished?

42. Why does a bypassed economiser need soot blowing as if in service?

43. What may have to be done if running with bypassed heat recovery devices?

44. What is the only cure for soot and hydrogen fires?

45. What is a minor furnace explosion called?

46. What operation is carried out to prevent furnace explosions?

47. What is done to prepare a boiler for an extended lay-up?

18.6 Questions – Tube Failures

1. What gives a warning of tube failure?

2. What is the difference between tangent wall and monowall construction?

3. What must be done if a membrane wall tube is plugged?

4. What kind of repair should be carried out on a membrane wall tube?

ANSWERS

1 FIREFIGHTING

1.1 Answers – Fire Prevention and Sources of Ignition on Ships

1. Examples of carelessly carried out maintenance where fires may occur are:

   - Sheaving on high pressure fuel pipes that are not correctly replaced after work
   - fuel filter covers that are not correctly torqued
   - lagging on exhaust manifolds that is not replaced or properly replaced after maintenance.

2. Safeguards against common causes of electrical fires are:

   - Testing and regular maintenance of equipment
   - regular testing of overload devices
   - checking that insulation is in good condition
   - checking connections are made properly.

3. Spontaneous combustion is caused:

   when cargos such as coal, hemp, copra, grain, etc are carried in a damp condition. The centre of these cargos will have very little ventilation to allow for cooling effect, therefore the natural heat generated can build up to such a degree that combustion takes place. Due to the restriction of oxygen, the cargo will only smolder until part of the cargo is removed, then admitting additional air will cause the cargo to burst into flames.

4. a) Information regarding the safe carriage of hazardous cargos/substances can be found in the IMDG code book which stands for the International Maritime Dangerous Goods code. Volumes of this code will be found on the bridge. Along with the IMDG code book is a supplement which gives you information such as emergency procedures, fire-fighting

techniques, medical first aid, packing, etc, for the various hazardous goods that are carried.
   b) Benzene
   i. This cargo should be stored in sealed containers which must be in good condition and will not allow escape of vapours or liquids.
   ii. Its hazardous properties are that it is colourless, clear characteristic odour, boiling point of 5°C, flashes below melting point of minus 11° C.
   iii. Fire-fighting techniques for this substance are that water spray, CO_2, dry chemical or foam extinguishers can be used. Self-contained breathing apparatus sets should be used, as prolonged exposure can have chronic effects.
   iv. Medical effects:

   - On contact will cause redness and irritation
   - is absorbed through skin
   - will cause nausea, headaches and vomiting
   - high exposure will cause unconsciousness and death.

   v. Treatments after physical contact:

   - On contact, contaminated clothing must be removed immediately
   - area contaminated should be washed for at least ten minutes with fresh water and, if there are signs of burns, wash for a further ten minutes
   - keep a close eye on the patient for at least twenty four hours
   - if burns are severe and extensive, shore side should be contacted for medical advice.

5. An example of a ship board welding operation is welding a bracket onto a bulkhead.

   - Check bulkhead is not adjacent to any sort of tanks such as fuel, lub oil

- check adjacent compartment for anything that may ignite from heat transfer, also around the welding area, remove anything combustible
- whilst welding, have someone standing by as a fire sentry, with a fire extinguisher, and also have them check adjacent compartments
- once the welding operation has ceased, check the job periodically until it has cooled down, up to two hours afterwards.

6. Fuel tank overflow system.

All tanks overflow to an overflow tank via a line with an observation glass. This line also incorporates a flow alarm. Fitted in the overflow tank is a level alarm which will be activated when the tank is a quarter full.

All tank vents are fitted so that oil cannot overflow onto deck or into machinery spaces which may lead to fires. The vent from the overflow tank is led onto deck and fitted with wire gauze diaphragms.

1.2 Answers – Inert Gas and Extinguishers

1. Sketch of funnel inert gas system.

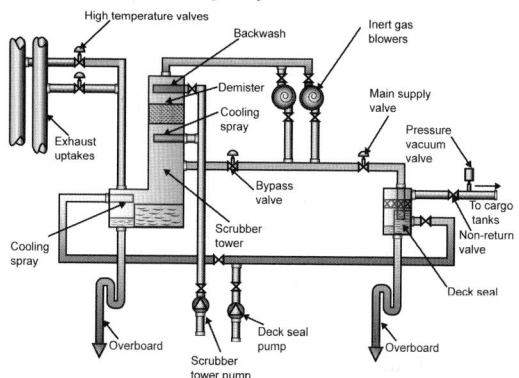

Engineering Pocket Book – Answers

The sketch shows an inert gas system from a main boiler. Gas from the boiler uptakes is taken via pneumatically operated high temperature valves, where it then passes through a scrubber tower.

In the scrubber tower, seawater is sprayed for cooling the gas to around 3 or 4° C above the seawater temperature, the seawater is also used to remove most of the soot and sulphur from the gas.

The gas then passes through a demister, which can be cleaned by means of back flushing.

After the scrubber tower, the gas analysis is as follows:

- Carbon dioxide @ 13%
- oxygen @ 3%
- sulphur monoxide 0.3%
- the rest nitrogen.

Two centrifugal fans then supply the dry, clean inert gas at 1.2 – 6 bar via a deck seal, non-return valve and pressure vacuum valve to the cargo tanks for inerting.

Safety features include.

- High oxygen content alarm 5% or above
- high gas temperature alarm
- low seawater pressure alarm (cooling)
- deck seal
- PV valve.

2. Which fire fighting equipment to use on certain fires:
   a) Small oil fire in the machinery space
   You could use a foam or dry powder fire extinguisher as this is a class B fire. These extinguishers would have a smothering type effect on the fire.
   b) Bedding fire in the accommodation
   You could use a water extinguisher on this type of fire as it is a class A fire. The water would have a cooling effect on the heat source.

Engineering Pocket Book – Answers

c) Galley fryer where it has been left on and the thermostat has failed, causing oil to burst into flames
As you don't know that the electrical supply has been isolated, you would use a CO_2 fire extinguisher on this type of fire. This would have a smothering effect on the fire.

3. a) Sketch of a portable foam extinguisher.

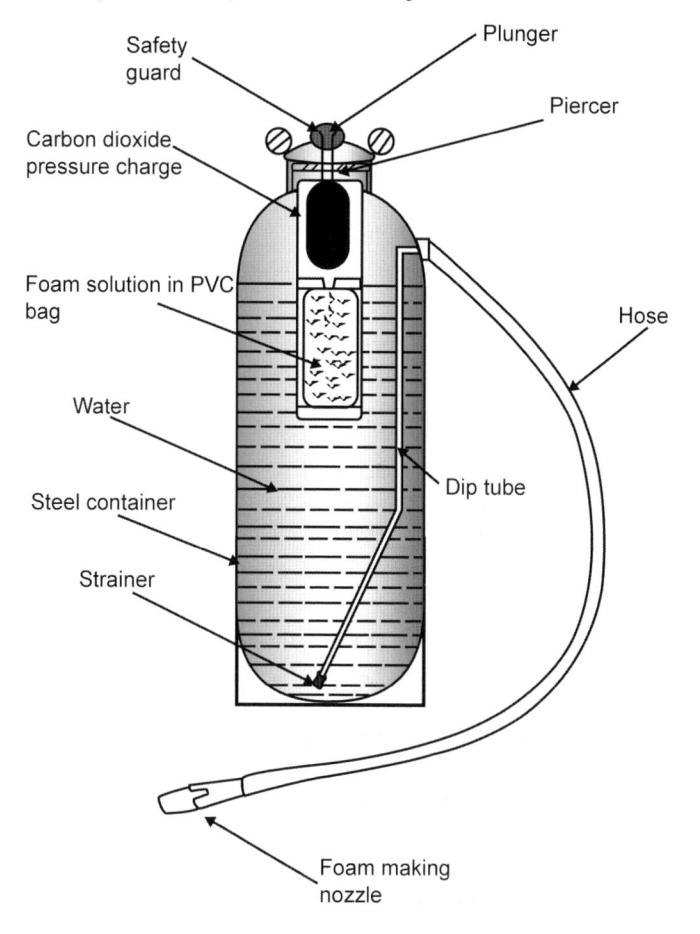

Labels:
- Safety guard
- Plunger
- Piercer
- Carbon dioxide pressure charge
- Foam solution in PVC bag
- Hose
- Water
- Dip tube
- Steel container
- Strainer
- Foam making nozzle

Engineering Pocket Book – Answers

When the plunger is depressed, it pierces the tin/copper seal, releasing CO_2 which ruptures the plastic bag containing foam solution and forces it to mix rapidly with the water. The foam solution is then driven up the dip tube through a hose to a nozzle, where it is aerated to produce good quality fire-fighting foam.

Performance:

- 9 L solution produces approximately 72 L foam
- jet length of approx 7 m for around 50 seconds
- can also be rapidly reloaded by filling with water and dropping new charge in.

b) Sketch of a dry powder extinguisher

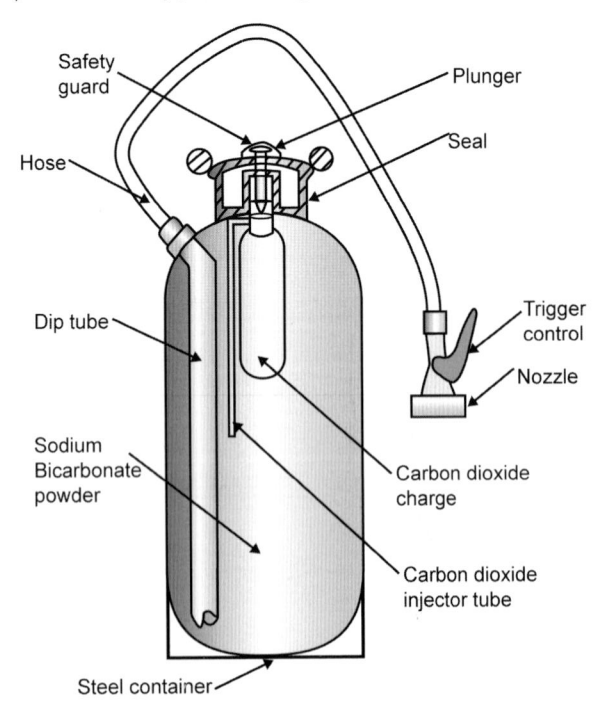

When the plunger is depressed, it pierces the CO_2 bottle seal, CO_2 then blows out the powder charge.

- Note the sodium bicarbonate has magnesium stearate added to stop the powder caking
- range of 3 – 4 m
- duration 15 seconds.

4. Flammable limits:

It is possible that an air/oil mixture can have too little or too great a content of flammable vapour and so will not burn mixtures between these limits are said to be within the flammable range.

1.3 Answers – CO_2, Sprinkler Systems and Detection

1. Advantages and disadvantages of the following:
a) CO_2 flooding system
   Advantages
   - Quick operation
   - rapid filling of space with CO_2
   - no power required to use the system.

   Disadvantages
   - Space has to be evacuated before use
   - does not sustain life, is deadly
   - fires can re-ignite after use
   - wait for a while before compartment is re-entered.

b) Water spray systems
   Advantages
   - Works automatically
   - safe to use
   - medium readily available
   - easy to test.

   Disadvantages
   - Damage after use (electrical equipment)

- has to be flushed through after use.

c) High expansion foam

   Advantages
   - Economic
   - rapidly produced
   - can use with existing ventilation system
   - little ill-effect on personnel.

   Disadvantages
   - Can take up to 48 hours to die down after use
   - large trunking required
   - if not trunked to bottom of compartment, convection currents can carry it away.

2. Sketch of a battery CO_2 system

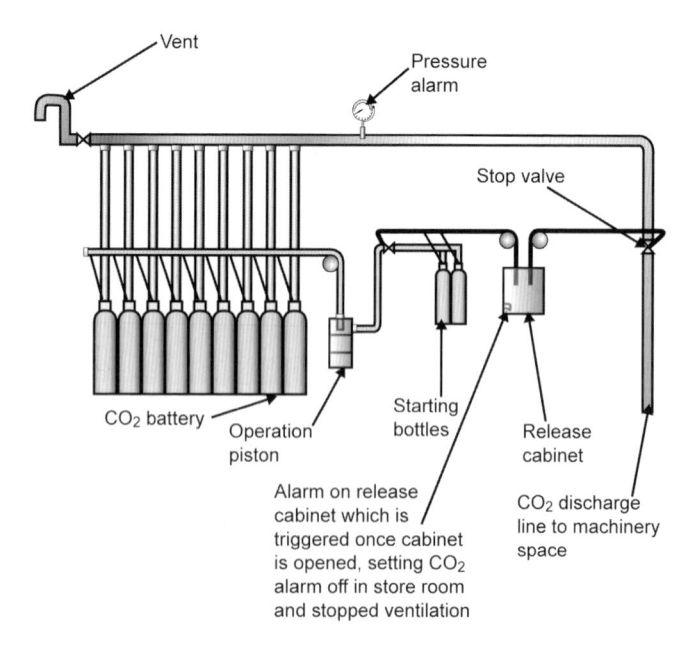

The sketch shows a CO_2 battery system for fire-fighting in a machinery space.

When the control cabinet is opened, an alarm is triggered which is audible and visual in the machinery space warning personnel that the release of CO_2 is imminent and that they should vacate the space immediately. Opening the cabinet will also stop the ventilation to the space.

The lever for releasing the CO_2 is then operated, which in turn operates the starting bottles. The gas from these bottles will drive a piston via a safety valve, and this piston releases the main battery of CO_2 bottles through a pulley system. The CO_2 discharges to the machinery space.

The system incorporates a stop valve on the discharge line and also a pressure alarm to indicate any leakage from the CO_2 battery.

The system must give 40% saturation of the whole compartment, in which 85% must be discharged into the compartment in the first two minutes.

3. The emergency fire pump on this ship is situated at the forward end of the ship (focsle), low down below the waterline so that it is always primed and ready for use. The fire pump has its own suction and is motor driven with its power supply from the emergency switchboard. The pump can be started locally, or from the bridge or from the emergency switchboard.

4. a) Fire patrols
   These are normally carried out on most ships, especially on sailing in hold compartments, engine rooms, boiler rooms, etc.
   They are also carried out in drydock of shipyards when personnel have just vacated the ship as they may have been

using oxy-acetylene burning or welding equipment and caused the beginning of a fire without realising it.

In addition, everyone onboard should be on the look out for causes of any fire.

b) Heat sensors

These are fixed temperature detectors that sense a sudden rise in temperature and set an alarm off. They should not give an alarm if the temperature rise is gradual, ie change of climate or heating going on.

These type of sensors are useful in dusty environments as the sensors are completely sealed, but they do not give off as early an alarm as other types of detectors.

c) Infrared flame detectors

This type of detector is set off from the flicker of flames. The detectors are tuned to go off at around 25Hz, which is the characteristic flicker of flames. They have a short time delay incorporated in the unit to minimise false alarms.

They can give an early warning of fire and this makes the detector suitable for areas where there is a high risk of fire, ie machinery spaces. They should not be installed in boiler rooms where naked flame torches are used for ignition.

d) Photo electric cell smoke detectors

There are three types of this detector in use, those that operate by light scatter, by light obscuration and a combination of both.

These types of detectors give a very early warning, but they can be vulnerable to vibration and dirt.

5. Testing of the above in situ:

- Heat sensors - these can be tested by means of a portable hot air blower
- infrared flame detectors - can be tested by means of a naked flame

- photo cell smoke detectors - can be tested by means of an aerosol can which has been specially formulated for testing these types of sensors, or from smoke from a cigarette.

1.4 Answers – Classification Regulations

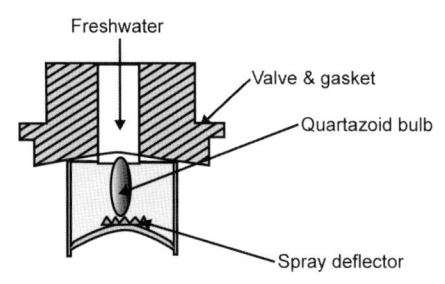

1. a) Sketch of sprinkler system suitable for accommodation spaces.

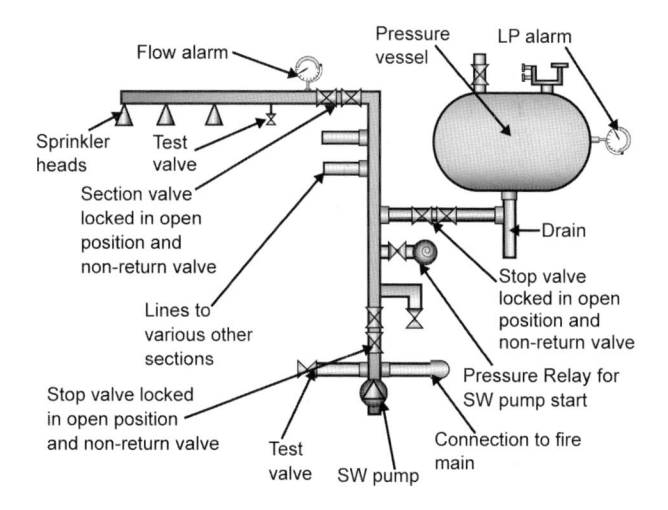

The sketch shows a sprinkler system suitable for accommodation purposes. A fire in a one section would cause a rise in temperature, causing a quartazoid bulb in a sprinkler head to break, which would then start spraying the area with fresh water. Once this happens, a flow alarm for this section will operate giving an alarm on a control panel that is normally situated on the bridge. This panel will show the section where the sprinkler head has activated, moulting in someone investigating the situation. If it is a genuine alarm, the fire alarm will be sounded, if a false alarm, the manual stop valve for this section, which will be locked open, can be shut, the key normally being behind a break glass panel in the vicinity of the valve.

The system is always pressurised with fresh water from a pneupress tank. In the event of a real fire and the pressure in the pneupress tank falls below a predetermined setting, a pressure relay will send a signal to the seawater pump, starting the pump and supplying seawater to the sprinkler system.

b) Advantages of the system are:
- Its automatic and quick operation
- the fire fighting medium is cheap and plentiful
- and the system is also easily tested.

Disadvantages are:
- It causes a lot of damage when used
- danger of free surface effect if a lot of water is used
- high up on the accommodation decks
- system has to be flushed with fresh water after use
- a good eye has to be kept on the pressure tank.

2. a) Sketch of a bulk CO_2 system.

Engineering Pocket Book – Answers

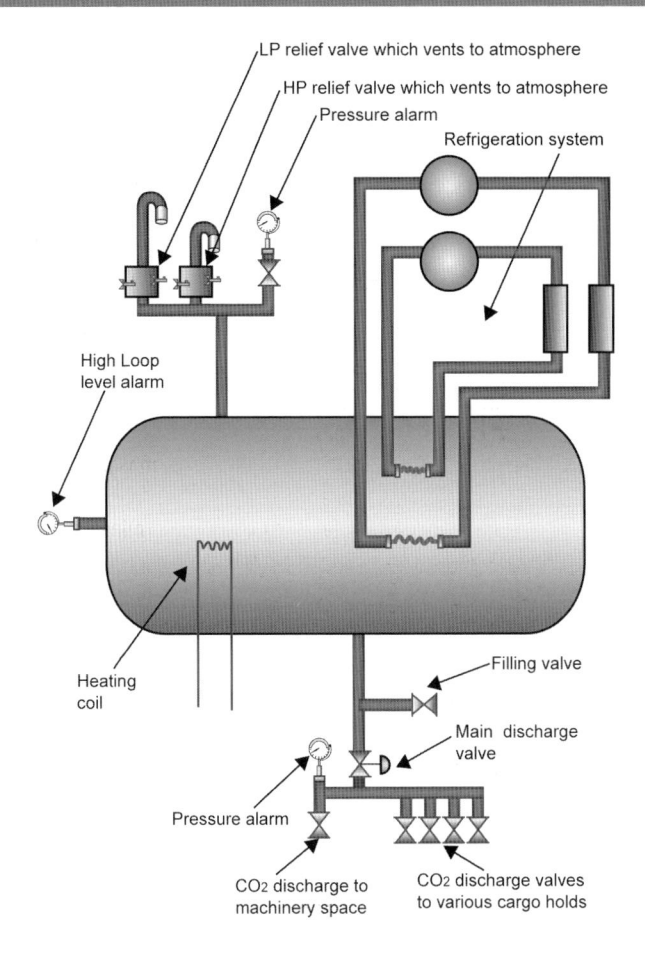

- LP relief valve which vents to atmosphere
- HP relief valve which vents to atmosphere
- Pressure alarm
- Refrigeration system
- High Loop level alarm
- Heating coil
- Filling valve
- Main discharge valve
- Pressure alarm
- CO2 discharge to machinery space
- CO2 discharge valves to various cargo holds

b) i. The system sketched consists of a large vessel which holds CO_2 at a working pressure of around 21 bar and temperature of minus 20°C. To maintain these parameters, a dual refrigeration system is employed and controlled by the CO_2 pressure in the vessel. One refrigeration system is in operation while the other is on standby.

125

As it is essential to maintain the pressure in the vessel, a heater is also fitted for this purpose. Fitted to the vessel are two sets of relief valves, the LP set designed to lift at around 24 bar and the HP set designed to lift at around 27 bar. The HP set is also required to vent off into the compartment. This is a safety feature employed in-case of fire in the compartment with the CO_2 vessel, causing a rapid rise in CO_2 pressure. CO_2 would vent off to the compartment, extinguishing the fire.

Alarms fitted to the system are low level, high level and a leakage/flow alarm (indicating leakage via main discharge valve). Balloons can also be fitted to the relief valves to give an indication of leakage.

To operate the system, the control/release cabinet is opened which sets off an audible and visual alarm, warning personnel that CO_2 release is imminent and that they should vacate the compartment. The section valve to the compartment to be smothered is opened, then the main CO_2 discharge valve is opened. A preset opening time is given for that compartment so that the correct charge of CO_2 is given.

ii. The advantages of a bulk CO_2 system over a multi-bottle system are that it gives a 50% less weight saving, the volume it occupies is less and it is a lot cheaper to supply CO_2 in bulk.

3. a) The of various means by which explosive gas may be inadvertently ignited by electrical machinery are:

- Bad insulation
- short circuits
- localised heating (motor bearings)

- overloads
- incorrect fittings (ie light fitting not explosion proof).

b) The probability of the ignition of explosive gas by electrical machinery may be reduced by:

- Regular electrical and mechanical maintenance being carried out, including insulation readings and cleanliness
- regular testing overload devices
- equipment being installed correctly and by a qualified person, and only used for that specific function.

4. Sketch of an International shore connection.

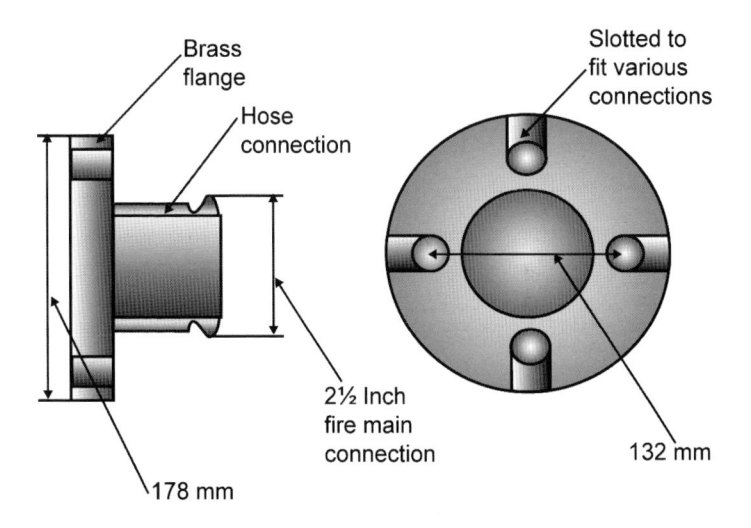

Brass flange
Hose connection
Slotted to fit various connections
2½ Inch fire main connection
132 mm
178 mm

Engineering Pocket Book – Answers

2 SAFETY

2.1 Answers – Enclosed Spaces and Entry

1. Oxygen deficiency may occur in certain spaces onboard ship due to:

   - Rusting which extracts oxygen from the air painting without sufficient ventilation
   - fuel tanks may also be deficient of oxygen due to hydro–carbons.

2. Precautions taken before entry into an enclosed space are as follows:

   - An entry permit must be obtained from a responsible officer, you may be the issuing officer
   - the space is to be well ventilated and oxygen levels tested by an O_2 meter
   - safety gear, such as self-contained breathing apparatus and rescue lines must be at hand
   - while anyone is in the enclosed space, someone must be on standby at the entrance who must be in communication with the people in the space and must be able to raise the alarm if they are to get into difficulties.

3. a) Sketch of BA set
   The sketch shows a self-contained breathing apparatus set, which consists of an air bottle of 120 bar attached to a frame and harness. The air bottle is attached to a cylinder stop valve and high pressure reducing valve which supplies air to the demand valve on the face mask via a low pressure alarm and pressure gauge. The demand valve on the face mask can be operated in two modes, constant pressure or on demand (ie as you breath). The duration for this type of set is 25 – 30 minutes, but this

Engineering Pocket Book – Answers

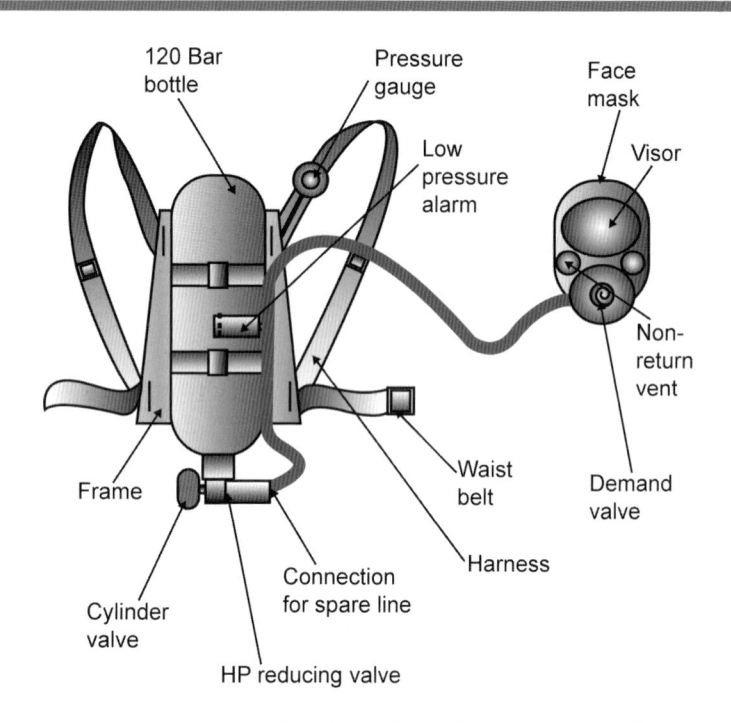

Diagram labels:
- 120 Bar bottle
- Pressure gauge
- Face mask
- Low pressure alarm
- Visor
- Non-return vent
- Frame
- Waist belt
- Demand valve
- Cylinder valve
- Connection for spare line
- Harness
- HP reducing valve

may be greatly reduced depending on the rate of physical work.

b) A bottle is checked before use to ensure it is fully charged and able to give its full duration of use.
Check carried out are:

- Bottle check – to ensure it is fully charged and able to give its full duration of use
- all hoses – to ensure they are okay and these are no lakes
- low pressure alarm – this will usually operate when there is 20% left in the bottle, allowing 5 minutes escape time
- face seal check – to ensure a good seal on the face mask.

4. Light gases may be found in the following tanks:

a) Permanent ballast tanks–hydrogen gases may be present in these tanks if cathodically protected.

b) Cargo tanks–If carrying oil you may find hydrocarbon gases present, but other cargos may give off various other gases such as carbon monoxide, sulphur dioxide, nitrate oxide, nitrogen dioxide.

c) Pump rooms–same as above, but most likely hydrocarbons, H_2S.

d) Fresh water tanks–methane gases may be present due to build up of micro organisms decaying at the bottom of the tank.

5. a) A description of a permit to work from.
A permit to work form normally consists of a description of the work to be done, who by, its period of validity and location. Most forms have check lists, ie for entering enclosed spaces, hot work or isolation of machinery or, if the form does not mention the work to be performed, there will be a column labeled OTHER. It may also have additional precautions.
The form is then signed by the person performing the task, and also by a responsible officer and is
Signed by both on the completion of work.

b) When a permit to work is required.
A permit will be required for entering cargo tanks or enclosed spaces, when hot work is performed, when electrical or mechanical isolation is required, or working aloft.

6. Toxicity of oil cargos or fuels.

Hydrocarbon gases are very toxic as well as being highly flammable and may be present in fuel oil or oil cargo tanks which have contained crude oil or its products.

Engineering Pocket Book – Answers

The components of some oil cargos vapours are lethal, such as benzene and hydrogen sulphide.

7. The safest way to gas free an oil tank which is fitted with an inert gas system is to use a fan and vent the tank. Before entry, an O_2 meter must be used and the tank must also be checked with an explosimeter. These meters must also be used while in the tank.

Sketch of an explosimeter.

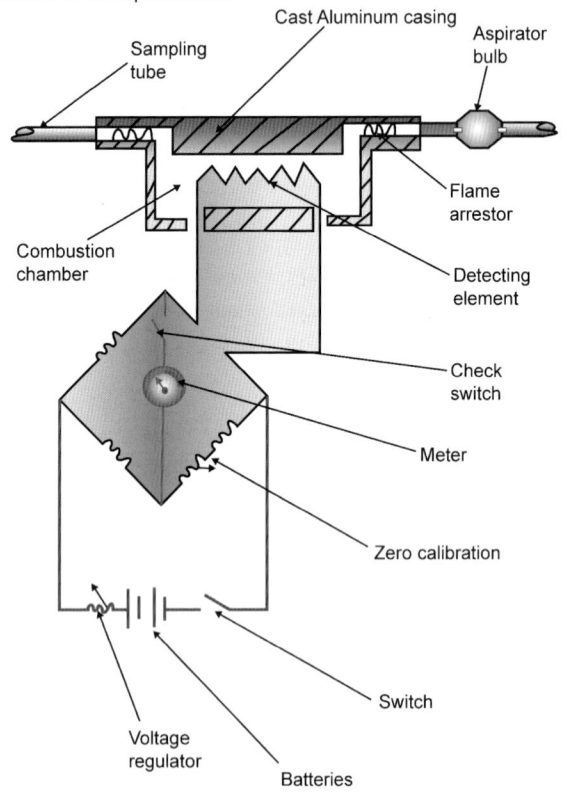

Labels: Cast Aluminum casing, Sampling tube, Aspirator bulb, Flame arrestor, Combustion chamber, Detecting element, Check switch, Meter, Zero calibration, Switch, Voltage regulator, Batteries

Engineering Pocket Book – Answers

The sketch shows a diagrammatic view of an explosimeter. In fresh air, the combustion chamber is cleared of any residual gas by the aspirator bulb. The meter is then switched on and time is allowed for the detector element to heat up. With the check switch closed, the zero adjustment of the meter is checked. In suspect atmosphere, the aspirator bulb is operated with the check switch closed, noting the meter reading. If there is gas in the atmosphere, this will burn in the combustion chamber due to the heating effect of the element. The burning will cause the temperature of the element to rise, giving a rise in electrical resistance of the element as it is proportional to the rise in temperature. This gives a reading on the meter which is calibrated to read the lower % of lower limit of explosive concentration of gas.

The O_2 meter works on the principle that oxygen is paramagnetic and that nitrogen is diamagnetic.

With reference to the sketch, two small spheres filled with nitrogen are arranged, dumb bell fashion, on a pivot between pole pieces. This causes a torque to be applied to the dumb bell.

The gas to be analysed is passed through the pole pieces. The magnetic field concentrates the oxygen in the gas and a small force, proportional to the oxygen concentration, exerts a further torque on the dumb bell arrangement.

The mirror on the dumb bell arrangement reflects light from a source to a photo cell where it is converted to an electric signal for input to an amplifier. The amplifier output is arranged to pass through a small coil around the dumb bell arrangement and generates a torque which is in opposition to that applied by the paramagnetic effect.

A current of measured flow is required to provide a balancing torque and this is proportional to the oxygen concentration.

Before entering the tank, the oxygen content must be 21%.

Engineering Pocket Book – Answers

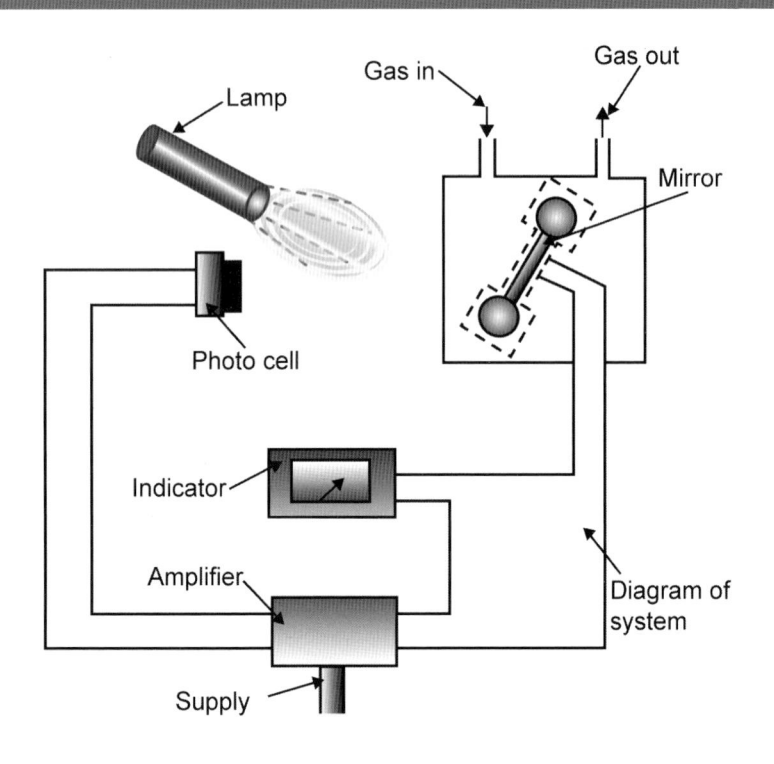

Recovery equipment must be prepared and left at the tank entrance, ie BA sets, lifeline and, depending on the type of tank, hoisting gear and resuscitator.

Communications must also be set up between personnel entering the tank and the person on standby at the entrance. He must also be in contact with the bridge to enable them to raise the alarm should a problem arise.

2.2 Answers – Lifting Plant and Welding

1. In the context of a ship's crane, the following terms are explained:

    a) Raising or lowering - this is where the hook is raised or lowered with the jib at rest.
    b) Slewing - this is where the crane jib is transversed horizontally with the hook at rest.
    c) Luffing - this where the crane jib is raised or lowered vertically with the hook at rest.

2. Precautions to be taken when working aloft:

    - Where possible, staging or a ladder should be used and secured firmly. A safety harness and lifeline must be used. If possible, a safety net should be rigged.
    - Tools should be carried up in secure containers or belts specially designed for carrying tools to prevent them falling and causing injury to anyone.
    - The area below the work site is to be roped off with warning signs starting, 'people working aloft.'
    - Care should also be taken when working aloft that tools are placed in secure positions to prevent them dropping on anyone.
    - Hazards should be identified, such as, ship's whistle, radars, antennas, which should all be isolated while aloft and 'do not operate' signs placed.

3. The precautions a duty engineer should take while working in the vicinity of the funnel are:

- To inform engine room
- to ensure steps are taken to reduce the emission of steam, harmful gases and fumes as much as possible.

2.3 Answers – Pollution and Harmful Substances

1. The dangers of asbestos dust are:

   - Asbestos dust can cause lung disease or lung cancer
   - where possible, any work involved which might generate asbestos dust in the atmosphere should be left until the ship is in port and proper facilities and equipment can be utilised
   - if it is essential for such work to be carried out, every precaution should be taken to generate as little asbestos dust as possible
   - access to areas where work is taking place should be limited to the personnel involved
   - where practical, working area should be closed off, ie, with plastic sheeting and signs
   - respiratory protective equipment approved for the purpose should be used.

2. a) Forces that contribute towards the total force available for oil and water separation are:

   - The force of gravity due to water and oil having different densities
   - oil will separate to the surface, water to the bottom
   - in the separator, heating coils, baffles, weirs and filters contribute towards separation.

b) Sketch of an oily water separator

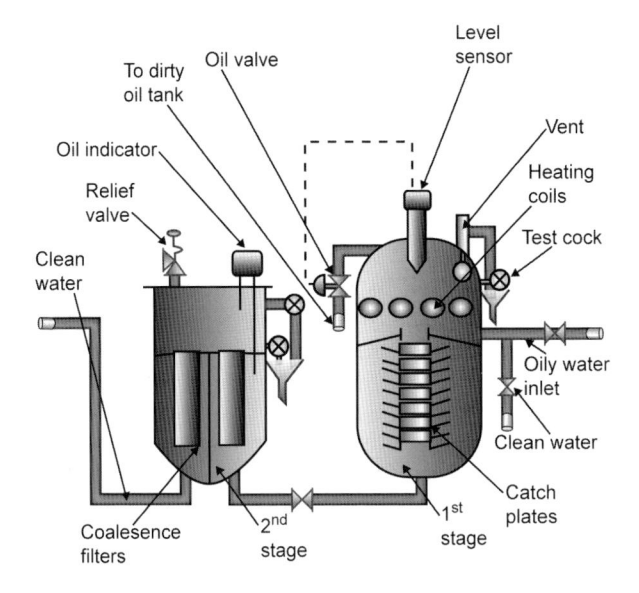

The complete unit is filled with clean water and then the oil/water is pumped to the 1st stage coarse separating compartment. Here, oil having a lower density than water, will rise to the surface with heating coils aiding in this process. This is known as the collection space. A sensor will then sense the level of oil and it will be dumped accordingly via an oil valve to the dirty oil tank. The remaining oil / water will move down to the fine separation compartment and move slowly between catch plates. More oil will separate on the underside of these plates and move outwards until free to rise up to the collection space. Almost oil free water then passes on to the 2nd stage of the unit. Here, two coalescer filters are situated, the first filter removes any physical impurities present and promotes some filtration, the second filter uses coalescer filter elements to achieve final filtration.

Clean water then leaves the second stage on to a clean water holding tank or via a 15ppm monitor with audible and visual alarms overboard.

c) Oil density and temperature relate to the ease of separation because with added heat, the viscosity of the oil is reduced and so aids separation, plus the density of the oil will lower and allow for better separation.

3. Sketch of a biological sewage treatment plant.

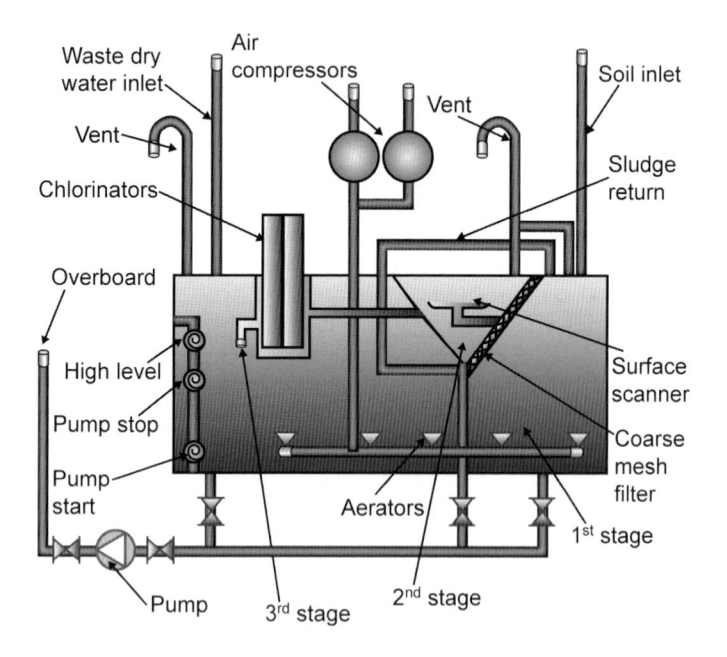

[Diagram showing a biological sewage treatment plant with the following labels: Waste dry water inlet, Air compressors, Vent, Soil inlet, Vent, Sludge return, Chlorinators, Overboard, High level, Pump stop, Pump start, Surface scanner, Coarse mesh filter, Pump, Aerators, 1st stage, 2nd stage, 3rd stage]

The sketch shows an aerobic sewage treatment plant. The system uses bacteria to totally break down the sewage for discharge overboard.

The system consists of three chambers. In the first chamber, aeration takes place, where sewage is broken down by aerobic bacteria, whose existence is aided by atmospheric oxygen which is supplied by compressors.

The sewage then passes through a coarse mesh filter to the settling chamber, where activated sludge settles out and returns to the aeration chamber.

Meanwhile, the clear fluid passes on to the final chamber via a chlorinator where any remaining bacteria is killed. In the final chamber, the clear fluid is automatically discharged overboard by a pump operated by float switches.

Four circumstances that may affect the quality of the output system are:

- Any oil or grease or the wrong type of cleaning fluids in the system will destroy the aerobic reaction
- unplanned shutdowns of the plant, as the plant has to be constantly in operation to maintain an efficient level of bacteria
- a build up of sludge will effect plant operation
- temperature change can also kill the bacteria.

The significance of Biological Oxygen Demand is the amount of oxygen taken up by the bacteria, and at the end of the process the sewage is said to be stable. As the activity of the bacteria reduces, so does the oxygen level and levels can be checked by taking samples and incubating them to check the effectiveness of the plant.

Colliform count:

This is a test taken to check the effectiveness of the disinfection process.

Results are given as the number of colliforms per 100 ml of effluent. One test involves incubating a sample over 24 hours at 35°C, another test involves incubating a sample over 24 hours at 35°C to see the colony of bacteria produced.

Engineering Pocket Book – Answers

4. The dangers of producing domestic water close to shorelines are:

- Toxic substances can be found in coastal water which contain pesticides, heavy metals, ammonia, cyanides, sulphides, fluorides, detergents, wastes from waterfronts and sewage effluent
- the domestic water treatment onboard ship will not be able to reduce the concentrations of toxicomania to a level fit for human consumption.

Sketch of an onboard water treatment.

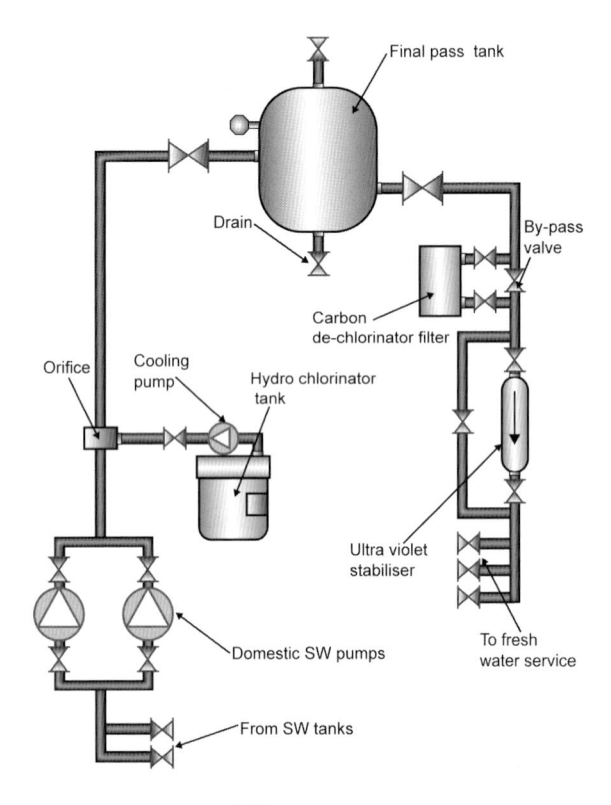

- Water is produced by the evaporator and pumped to the fresh water storage tanks
- the pneupress pumps take suction to pressurise the pneupress tank
- from the pneupress tanks, the fresh water passes through the hypochlorinator where water is sterilised by an excess dose of chlorine provided as hypochroinate solution or tablets
- it is then de-chlorinated in a bed of activated carbon to remove excess chlorine
- any colour, taste and odour which was present in the water will also be removed by the carbon
- the water is then passed through an ultra violet steriliser before entering the domestic system.

3 MATERIALS

1. Phosphorous is undesirable in steel because it appears as a hard, brittle constituent and, for this reason, it should not be allowed to exceed 0.05% content in steel.

2. The percentage of which steel becomes an alloy is 1% above manganese content.

3. The benefit of adding manganese to steel is that it de-oxidises the steel, ensuring freedom from blow holes.

4. The effect of temperature on grain boundaries is that at high temperatures, the grain boundaries are weaker and at relatively low temperatures, the grain boundaries are stronger.

5. The factors of recrystallisation of steel depend on:

    - The degree of prior cold working
    - the addition of other elements
    - the annealing time.

6. The difference between hot working and cold working is that hot working is done at temperatures above recrystallisation and does not produce residual stresses, where as cold working destroys the lattice structure and the metal then becomes harder and ductillity is lost.

    Hardening due to cold working is known as work hardening.

7. The percentage of carbon in the following steels are:

    a) Dead mild steel 0.07 – 0.15%
    b) mild steel 0.15 – 0.3%
    c) medium carbon steel 0.3 – 0.6%
    d) high carbon steel 0.6 – 1.4%.

Engineering Pocket Book – Answers

8. Advantages and disadvantages for the use of the above steels are:

- Dead mild steel is soft and ductile and can be used for rod and wire nails and rivets. It does not machine well and its tensile strength is only up to 350 N/mm^2 and elongation of 32%
- mild steel is used for ships plating, boiler plate, steel structural sections, it welds easily and has good machining properties. Its tensile strength ranges up to 450 N/mm^2 and elongation of up to 25%
- medium carbon steel is used for gearing, axles, crankshafts, it is harder and stronger than mild steel but less ductile. It is more difficult to weld and machine than mild steel. It has a tensile strength of up to 700 N/mm^2 and elongation of up to 12%.
- high carbon steel is harder and less ductile than medium carbon steels and is almost always fully heat treated before use. Properties vary with carbon content and methods of heat treatment but, in general, the lower the carbon content, the tougher the steel, the higher the carbon content, the harder and less shock resistant the steel.

9. The most generally used steel is mild steel because it is easy to weld, machine and it is inexpensive compared to most other steels.

10. The purpose of heat treating steels is to change the mechanical properties of plain carbon steel.

(Steel's upper critical temperature range is 700 – 900°C depending on carbon content)

11. Three types of heat treatment are:

Hardening

This involves heating the steel until it is 30 – 50°C above upper critical temperature range (850 – 950°C), followed by rapid quenching. The hardest possible condition for the steel is produced this way and the tensile strength is increased.

Tempering

This is employed to relieve quenching stresses and involves reheating the steel to around 650°C (lower critical temperature range). Note the higher the tempering temperature, the lower the tensile properties of the material. Once tempered, the metal is rapidly cooled by quenching.

Annealing and Normalising

Normalising is where the steel is heated to around 850 – 950°C depending on its carbon content and is then allowed to cool in air. A hard strong steel with refined grain structure is produced.

Annealing is where the steel is heated to around 850 – 950°C but is cooled slowly, either in a furnace or an insulated space. A softer more ductile steel than that of normalized condition is produced.

12. The process of nitriding is where nitrogen instead of carbon is used as the hardening agent. The process known as nitriding involves placing the steel component in a gas tight container through which ammonia gas is circulated. The container and component are then raised in temperature to approx 500°C and, nitrides are formed in the material close to its surface, which increases surface hardness.

Two examples where this is used is on crankshafts and camshafts.

13. Alloy steel is necessary to overcome limitations of plain carbon steels such as poor resistance to oxidation, corrosion and creep at high temperatures. Note that high tensile strengths cannot be combined with toughness and ductility.

Engineering Pocket Book – Answers

14. Austenite stainless steels would be used in the manufacture of high temperature valve spindles such as exhaust valves, exhaust gas turbo blowers and turbine nozzles.

15. The benefits of using titanium alloys are that they are high in tensile strength and have relatively low density, ie very good in turbine engineering, and they have good corrosion resistance.

16. The nimonic series of metals are not strictly speaking steels but contain 75% nickel and 20% chromium, stiffened with small amounts of carbon, titanium, aluminium, colbalt and molyadium. They are used where very high creep strength at high temperatures is required. (eg steam, gas turbines)

17. Stellite is the trade name of an alloy consisting of colbalt, chromium, molyadium, tungsten and iron. Stellite is extremely hard, corrosion resistant and has good resistance to loss of strength at high temperatures. It is used in diesel engines to surface the face of seats for exhaust valves.

18. Four alloying elements are:

Nickel, 1 – 8% used. The properties given to steel from nickel are that it increases strength and toughness with little loss in ductility, and has good corrosion resistance.

Chromium, 0.25 – 18% used. This is used in small amounts in constructional steels, tool steels and ball races.

It is used in large amounts for heat resisting steels and stainless steels. It induces hardness and improves resistance to erosion.

Vanadium, 0.2 – 0.4 % used. This increases strength and fatigue resistance. It is used in steels required to retain hardness at high temperatures, ie hot forging dies.

Mananese, 1 – 2 % used. Reduces the ill effect of oxygen and sulpher and it increases strength.

19. Fatigue in relation to metals is caused by repeated stress cycles such as reverse or alternating stresses, repeated stresses and fluctuating stresses.

20. Factors that influence fatigue strength of a material are:

- Grain size – when a material has a course grain it has poorer fatigue resistance
- component shape – sharp corners, keyways, etc
- surface finish – tooling marks act as stress raisers, so fatigue resistance of a component can be improved by polishing
- residual stresses – stresses produced by machining will affect fatigue resistance of a component
- corrosion – corrosion produces a pitted surface and so introduces stress raisers
- temperature – at high temperatures, materials tend to loose their strength and also suffer grain growth.

21. Creep is the slow plastic deformation that occurs under prolonged loading, usually at high temperatures.

22. Factors that influence creep are:

- Temperature – the rate of creep increases with temperature rise
- grain size – a course grain material has a better creep resistance than one with a fine grain size.

23. The constituents of cast iron are generally any material made up primarily of iron with about 2% or more carbon.

24. Four properties of cast iron are:

- Low melting point
- good wear resistance
- high compressive strength
- rigidity.

25. White cast iron is undesirable because it is hard, brittle and unmachinable.

26. The factors that affect the form taken by carbon in cast iron are the rate of cooling and chemical compositions.

- If cooling is too rapid, it tends to produce cementite and white cast iron, while slow cooling allows the precipitation of graphite in grey cast iron
- silicon aids the formation of graphite and is used for this process of producing a soft cast iron
- sulphur stabilised cementitic, making the iron hard and brittle, has its effects suppressed by the addition of manganese
- when high fluidity is required, phosphorous is used but the iron produced is weak and brittle but is useful for ornamental purposes.

27. Spheroidal graphite cast iron

Is cast iron where castings with the graphite in globular form can be obtained by adding small amounts of magnesium to the ladle before casting.

The metal is then annealed giving a structure consisting of globules of carbon in a ferrite matrix. Such castings can replace steel castings and forgings.

28. Copper would be used for electrical applications such as wiring, connections and domestic pipework. Its greatest use is in alloying with non-ferrous metals in the production of bronze, gunmetal, cupro-nickel and other alloys.

It has good mechanical properties such as being soft and ductile, easy to shape and with good corrosion resistance when in contact with water or steam.

29. Dezincification of brass is the removal of zinc content by corrosion by seawater or hot water, leaving behind a porous and leaf-like coloured sponge copper.

30. Dezincification of brass is prevented by adding a small amount of arsenic to the brass.

31. The properties bearing metals should have are:

- A low coefficient of friction
- be sufficiently hard to resist wear
- be tough to withstand shock loading
- have strength to support the working load
- have sufficient plastic to allow self alignment
- have high thermal conductivity to dissipate heat when running.

32. Sintered bearings are made by heating a compressed powder mixture of 90% copper and 10% tin with an addition of graphite.

These bearings are semi-porous and retain lubricant.

33. Electrochemical corrosion is a term that covers all forms of wet corrosion, ie where the metal is in contact with a liquid or even a moist atmosphere. In electrochemical theory, it is assumed that all metals have a tendency to dissolve or corrode, this is when the metal discharges positively charged particles called ions into solution. This leaves the metal with a characteristic negative potential, the greater is the tendency of the metal to dissolve or corrode.

34. The corrosion is most severe when in contact with a dissimilar metal in a conducting liquid known as an electrolyte.

It also depends on relative areas of the anode and cathode in the periodic table.

35. The anode is the negative and the metal attacked by corrosion, higher in table.

The cathode is the positive and protected, depending on the given area in the table, the severity of the attack is greater if metals are further apart.

36. Pitting is an example of the differential aeration effect. The initial depression or pit in the surface may be a result of several factors, ie a break in the protective film or scale. Once a pit is formed, corrosion proceeds rapidly since the surface of the metal (cathode) has a greater access to oxygen than the base of the pit (anode).

37. TIG welding is welding where an arc is drawn between a (tungsten inert gas) water cooled non-consumable tungsten electrode and a plate. An inert gas shield is provided to protect the weld from the atmosphere (argon gas) and filler metal may be added to the weld pool as required. Ignition is obtained by means of high frequency discharge across the gap.

38. Compressed asbestos fibre jointing is used when severe physical conditions are attained such as in use in diesel exhaust manifolds, steam lines and some cylinder head joints where vibration is a problem.

39 & 40. Advantages of PTFE as a jointing materials are that it can replace materials such as asbestos, glass, ceramic and rubber in some applications.

It has a high chemical resistance except against molten alkali metals and elemental fluorine particulary at high temperatures.

PH range 0 – 14 Operating pressure, vacuum to 240 bar.

41. Factors that effect sealing characteristics are that the gasket needs to be compressible within contours resolved. High temperatures also effect seals as there is a reduction in force applied by clamping bolts, the gasket may also undergo chemical charges due to heat. These processes must not effect or reduce the pressure between the gasket and the surface to which it is sealing as the gasket can be damaged by the internal pressure.

4 AUXILIARY MACHINERY

4.1 Answers – Heat Exchangers

Answers

1. Points to be considered in the design / selection of a heat exchanger are:

   - The quantity of fluid to be cooled, min – max
   - range of inlet and outlet temperatures of fluid being cooled
   - as above for cooling medium
   - specific heat of mediums
   - type of medium, ie corrosive or non-corrosive
   - operating pressures
   - position of system.
   - cost, materials, streamline or turbulent flow.

2. Streamline flow:

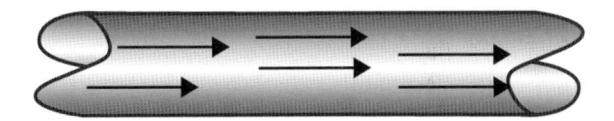

The drawing above shows streamline flow of a liquid whose velocity variation is parabolic, being maximum at the centre and zero where fluid is in contact with the pipe or plate surface.

Turbulent flow:

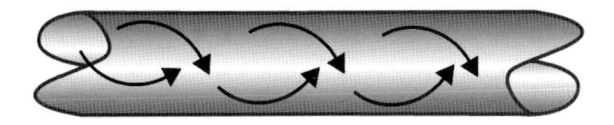

Engineering Pocket Book – Answers

The drawing above shows turbulent flow of liquid where velocity is present even at the pipe surface. For efficient heat transfer, turbulent flow is best, but erosion of metal surface is greatest.

For little erosion, streamline flow is required, but heat transfer will be relatively poor.

Whether flow is streamline or turbulent depends on certain factors which are summed up by the REYNOLD'S number:

- If the number is less than 2000, the flow is streamline
- if the number is greater than 2000, the flow is turbulent.

3. Sketch of a shell type heat exchanger.

The sketch shows a tube/shell type cooler that could be used as a main engine jacket water cooler. The cooler shown is of the two pass type, the jacket water being cooled, coming in contact with the outer side of the tube stack, and the shell surface. Baffles are used to direct the flow of jacket water through the cooler, and also support the tube stack.

Seawater, being the cooling medium, is in contact with inside of the tubes.

Materials used in the construction of the cooler are as follows:

- Shell – cast iron or fabricated steel
- waterboxes – fabricated from mild steel, their insides coated with rubber or anodes placed for protection against corrosion
- tubes – these are made from aluminium brass or cupro nickel
- tube plate – made from naval brass
- to allow for thermal expansion, one end of the stack is fixed, while the other is free to move to allow for expansion.

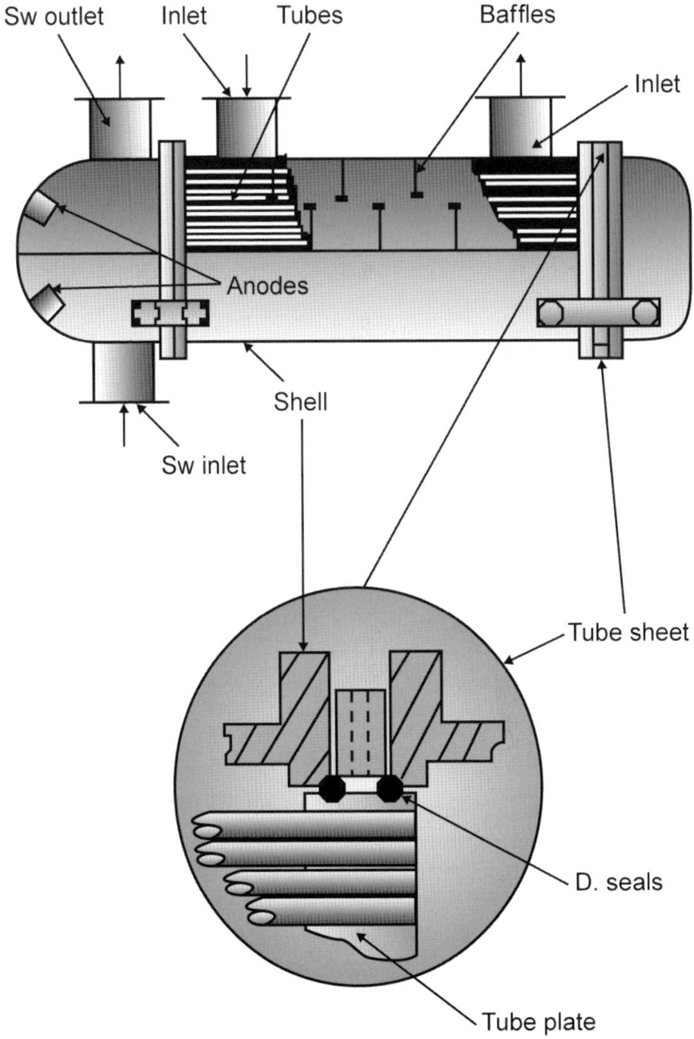
Sw outlet
Inlet
Tubes
Baffles
Inlet
Anodes
Shell
Sw inlet
Tube sheet
D. seals
Tube plate

Engineering Pocket Book – Answers

4. Plate type heat exchanger.

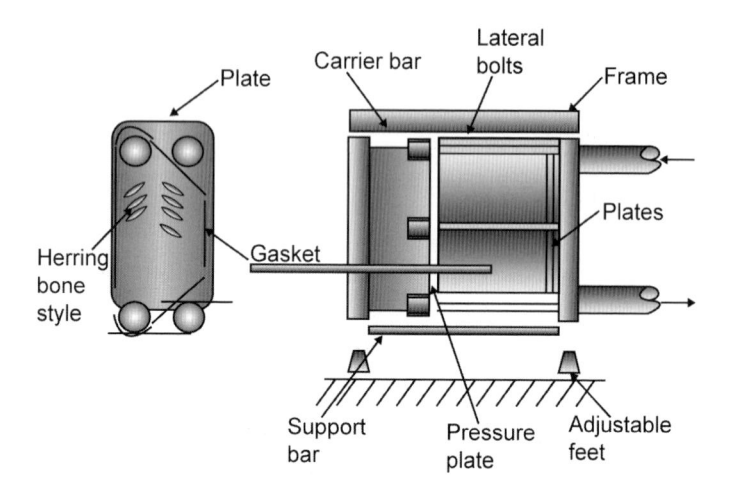

This sketch shows a plate type heat exchanger. It consists of a variable number of gaskets and plates, clamped together between a frame and a pressure plate. The surface of the plates are corrugated in a herring bone style to give extra strength and a larger cooling surface area.

5. The principle advantages of a plate type heat exchanger are:

- Compact and space saving
- easily inspected and cleaned, all pipe connections are at the frame plate, so they do not need to be dismantled when plates are removed
- variable capacity, plates can be altered to meet capacity demands
- with titanium plates, there is virtually no erosion risk and they allow for turbulent flow which takes place between the plates to give greater heat transfer and enable fewer plates.

6. The function of a condenser is to remove latent heat from exhaust steam so that the condensate obtained can be handled by pumps in the feed system.

It is also required to reduce the back pressure and so allowing a greater amount of work to be done by the engine, improving efficiency.

7. The two principle types of condenser are:

- Surface – in the surface type, there is a temperature drop of about 8°C from inlet to outlet and the condensate and air leave from the bottom
- regenerate – in this type, steam flowing along the regenerative passage and up into the tube nest heats the condensed droplets from the tubes so there is practically no temperature drop in the condenser. Air trapped by the exhaust baffles is extracted separately by an ejector.

8. An 'End on' or axial exhaust condenser:

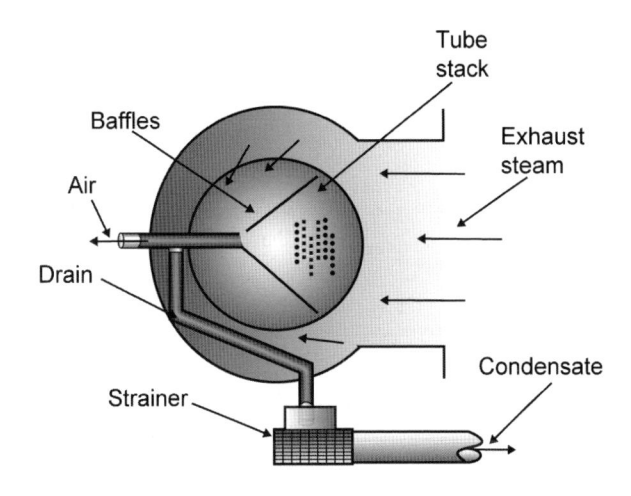

Exhaust steam enters the condenser side on, this reduces LP turbine casing distortion and since the exhaust steam leaves the LP turbine straight into the condenser, steam losses are reduced compared to under-slung types.

9. Materials used in condenser construction are:

- Shell – mild steel or cast iron
- waterboxes – cast iron
- tube plates – brass
- diaphram and baffle plates – mild steel
- tubes – aluminium bronze, stainless steel, cupro-nickel
- stays – mild steel with stainless steel cap nuts.

10. Tubes are fixed in a condenser by:

- Being packed at both ends
- expanded at the inlet and packed at the other
- expanded at both ends.

11. A defective condenser tube can be located by:

Method one

Empty seawater side and open up condenser, then place light polythene over tube plate and create a vacuum. Suction will indicate faulty tube or tubes.

Method two

Fill steam side with water and add flourescein powder to the water. By shining ultraviolet light on the seawater side of the condenser tube plate, the faulty tube can be located by bright green drips.

4.2 Answers – Pumps

1. Passage of water through a centrifugal pump.

    Water enters the pump axially through the eye, then by centrifugal action continues radially and discharges around the entire circumference. The fluid, in passing through the impeller, receives energy from the vanes giving an increase in pressure and velocity. The velocity, which is kinetic energy, is partly converted into pressure energy by suitable design of the impeller vanes and the pumps volute casing.

2. Sketch of centrifugal pump.

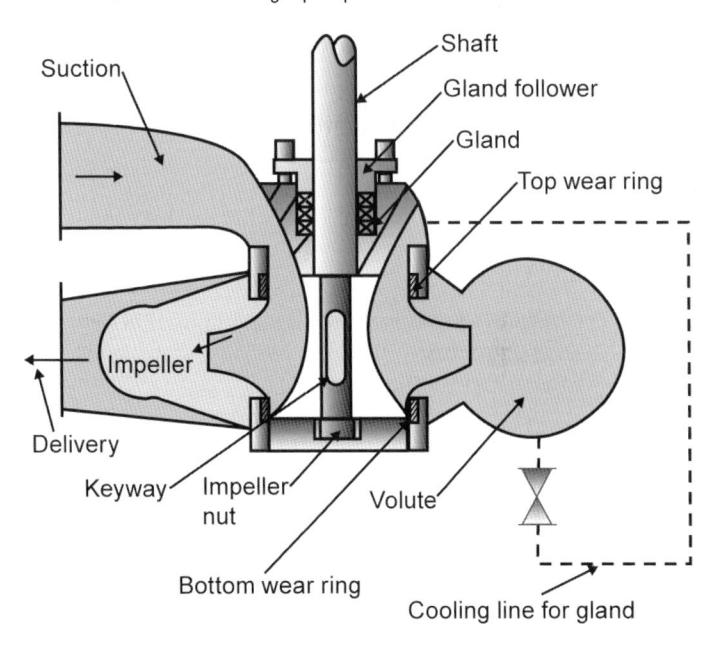

Sketch of impeller and volute casing.

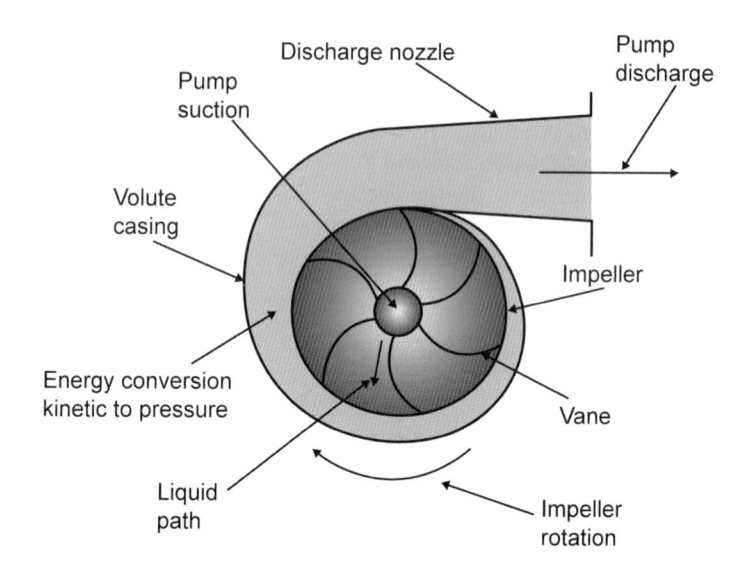

Labels: Discharge nozzle, Pump discharge, Pump suction, Volute casing, Impeller, Energy conversion kinetic to pressure, Vane, Liquid path, Impeller rotation

The centrifugal pump consists of a stainless steel shaft with keyway for the impeller, which is normally made of aluminium bronze and fitted by means of a bolt to the shaft. Replaceable wear rings/sealing rings are fitted to the pump casing and impeller, these separate the suction from the discharge side of the pump.

The pump casing is normally made of gunmetal or cast iron but will depend on the pump's application.

The sealing arrangement can be by means of gland packing or mechanical seal.

3. Positive displacement pumps are pumps in which one or two chambers are filled and emptied. These include reciprocating, screw, gear and water ring type pumps.

Engineering Pocket Book – Answers

They do not require a priming device and may themselves be used as one.

They cause a reduction and increase in volume of space and cause the liquid or gas to be physically moved.

4. Dynamic pressure pumps are pumps in which a tangential acceleration is imparted on the fluid. These include centrifugal flow, axial flow, and mixed flow types (the latter is a combination of centrifugal and axial flow), depending on the head they may require a priming device.

5. Sketch of an axial flow pump.

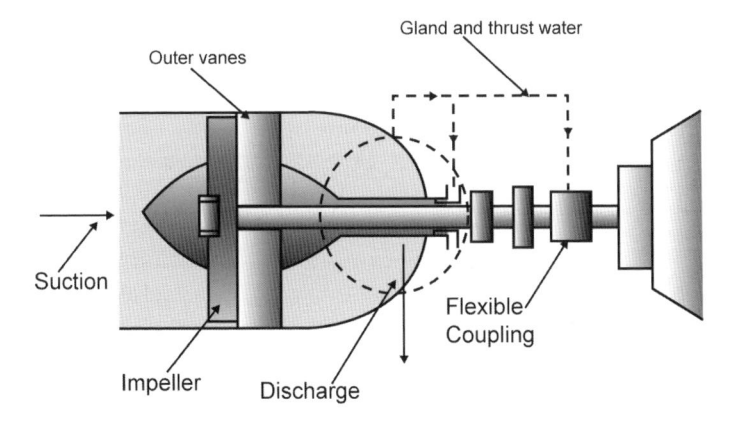

6. Sketch of a gear pump.

In the gear pump sketched, fluid enters the suction side, drawn in by two gear wheels that are meshed together and are close fitting to the pump casing, and is carried round between the teeth of the gear and the casing.

Such pumps are fairly efficient and smooth running and are best suited to pumping oils, ie good in fuel systems.

Engineering Pocket Book – Answers

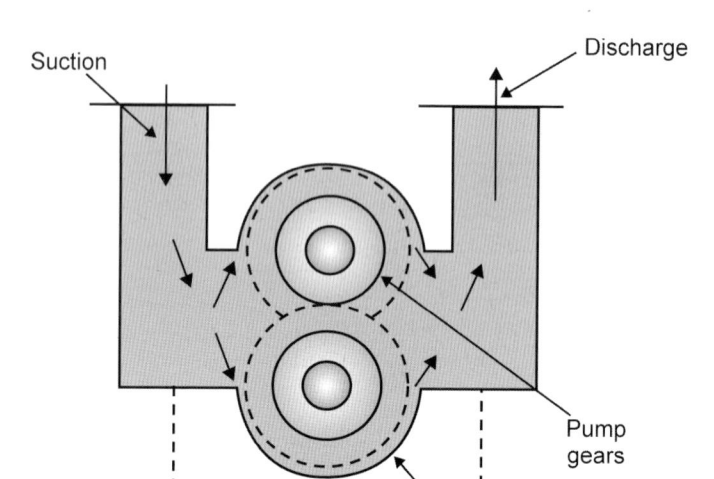

Note: these pumps must also be fitted with a relief valve.

Diagram: Gear pump showing Suction, Discharge, Pump gears, Pump casing, and Relief valve.

7. Sketch of a piston pump.

   In the pump sketched below, fluid is entering the pump on the suction side when the piston is on the upward stroke, the bottom suction valve is open and the bottom discharge valve is closed, drawing in fluid at the bottom. At the same time, the discharge valve at the top is open and the top suction valve is closed, discharging water at the top and vice versa when the piston is on the downward stroke, meaning the piston is discharging and sucking on both strokes.

   The air vessel in the system is designed to take up any fluctuations in pressure.

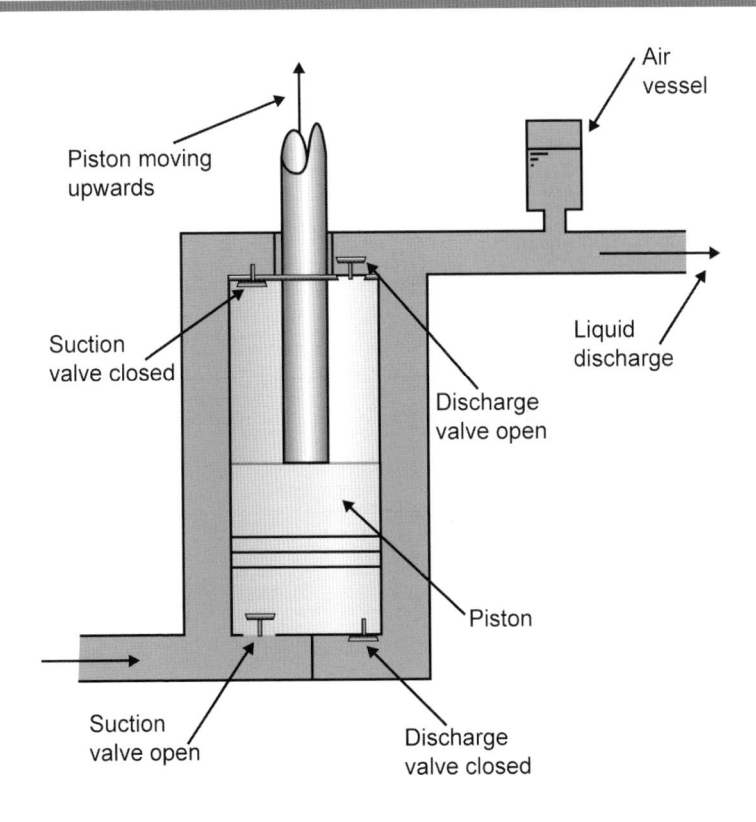

Figure labels: Piston moving upwards; Air vessel; Suction valve closed; Discharge valve open; Liquid discharge; Piston; Suction valve open; Discharge valve closed

8. The type of pump that I would choose for an emergency bilge pump would be a centrifugal pump with a self-priming device, as centrifugal pumps can pump high volumes of water (60 kg/s), and they also require little attention.

   Some emergency bilge pumps have their motor encased in an air bell so the pump can be operated when submerged.

Engineering Pocket Book – Answers

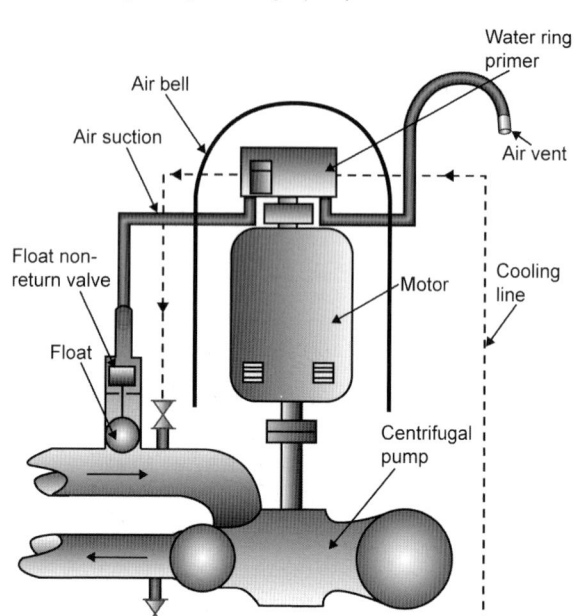

Sketch of self-priming centrifugal pump with air bell.

9. Cavitation is very high pressure hammer blows caused by pressure regions occurring in the fluid flow at points where high local velocity exist. If vaporisation occurs due to low pressure areas, then bubbles occur, these expand as they move with the flow and collapse when they reach a high pressure region caused by high pressure hammer blows. This can lead to pitting, noise, vibration and fall in pump efficiency.

To prevent this, some centrifugal pumps have inducers fitted to the pump impeller and their purpose is to ensure there is sufficient supply of fluid at the impeller, avoiding cavitation at impeller suction. They will also allow the pump to operate with a lower net positive supply head.

Different types of inducer are used, either scroll, screw or impeller.

4.3 Answers – Filters

1. Sketch of an Auto-klean filter.

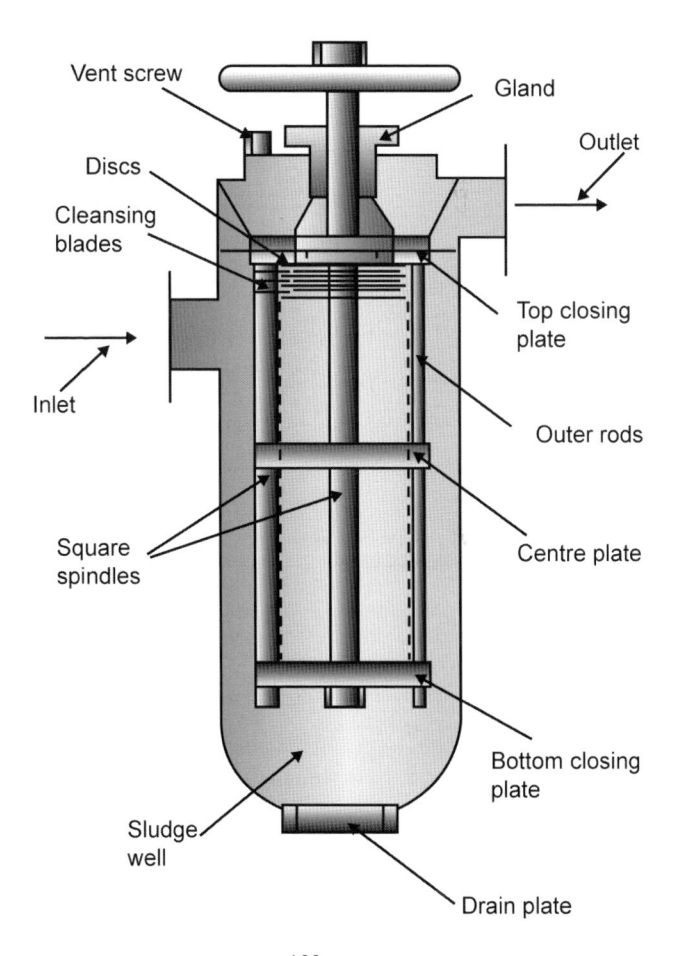

Engineering Pocket Book – Answers

This filter can be cleaned while in operation. The filter is cleaned by rotating the centre square spindle which rotates the disc stack and the stationary cleaning blades scrape off filtered solids which settle to the bottom of the sludge well of the filter.

Periodically, the flow of oil is disrupted and the sludge well is cleaned out. To facilitate this, the filters are generally fitted in pairs.

2. Sketch of streamline lubricating oil filter.

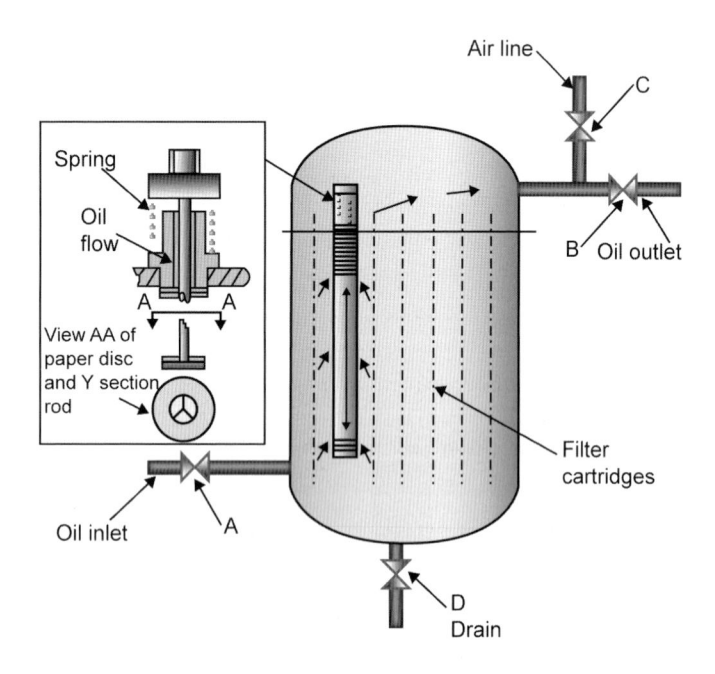

The streamline lubricating oil filter consists of two compartment pressure vessels containing filter cartridges. Each cartridge is made up of thin discs threaded on to a 'Y' or 'X' shaped section rod and held in compression. The oil flows from

Engineering Pocket Book – Answers

the dirty to the clean side of the filter via small spaces between the compressed discs, then up the spaces formed by the poles in the centre. In this way, dirt is left at the edge of the disc stack.

It is claimed particles of 1 micron can be filtered out.

For cleaning, generally compressed air is used, valves A and B are closed and D and C are opened and reversal of flow results.

3. Coalescing action.

Coalescing action filters, normally consist of some form of prefilter to remove particles, then followed by a compressed inorganic fibre coalescing unit in which water is collected into larger globules.

Coalescing action is the molecular attraction between the water droplets and the inorganic fibres is greater that that between the oil and the fibres, when the water globules are large enough they will move with a stream out of the coalesing unit.

Downstream of the coalescing cartridges are PTFE covered stainless steel water repelling screens which act as a final barrier for the water. Water gravitates from them and from the outlet of the coalescer cartridges into the well of the filter body from where they are periodically removed.

4. The size of particle an auto-klean filter can filter out is in order of 75 microns, but a more recent type can filter out down to 25 microns.

5. Pressure gauges are fitted to filters before and after the filter to determine the state of the filter, ie dirty or not.

6. Sketch of a lubricating oil coalescer filter.

The coalescer filter may be cleaned by draining the water off manually every so often or it may have its own back flush system.

Engineering Pocket Book – Answers

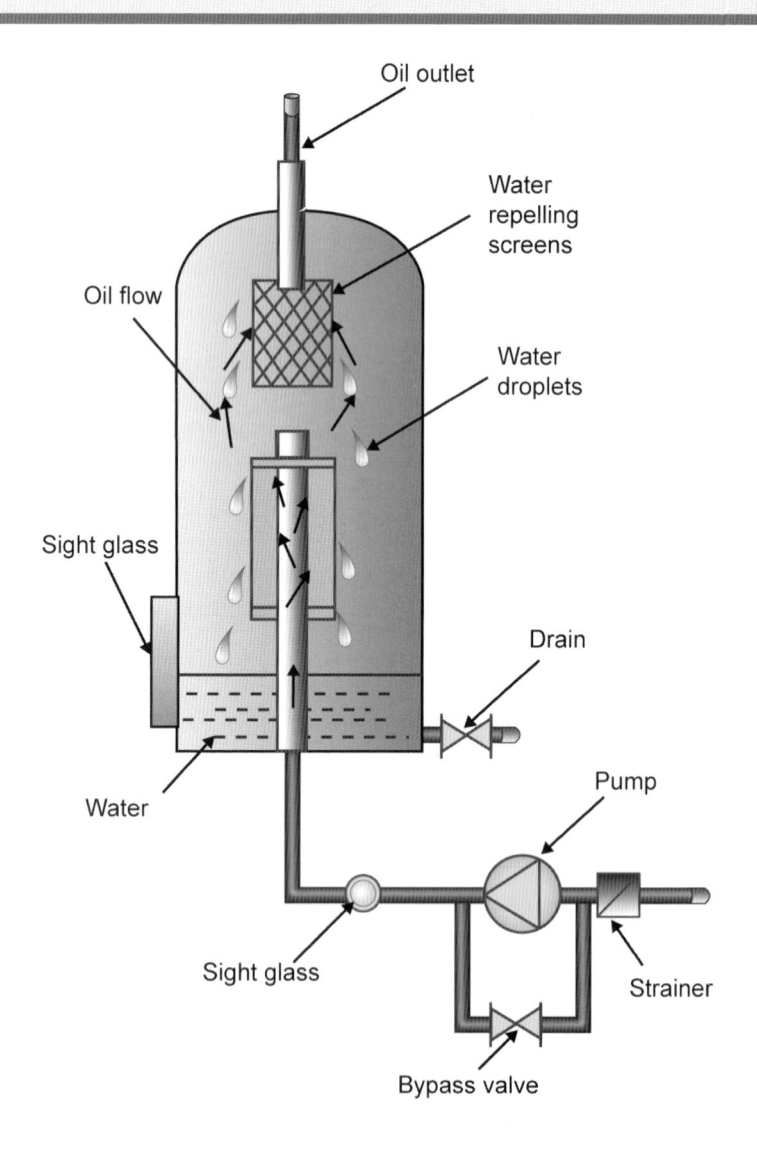

Engineering Pocket Book – Answers

7. Sketch of back flush system for an automatic oil filter module.

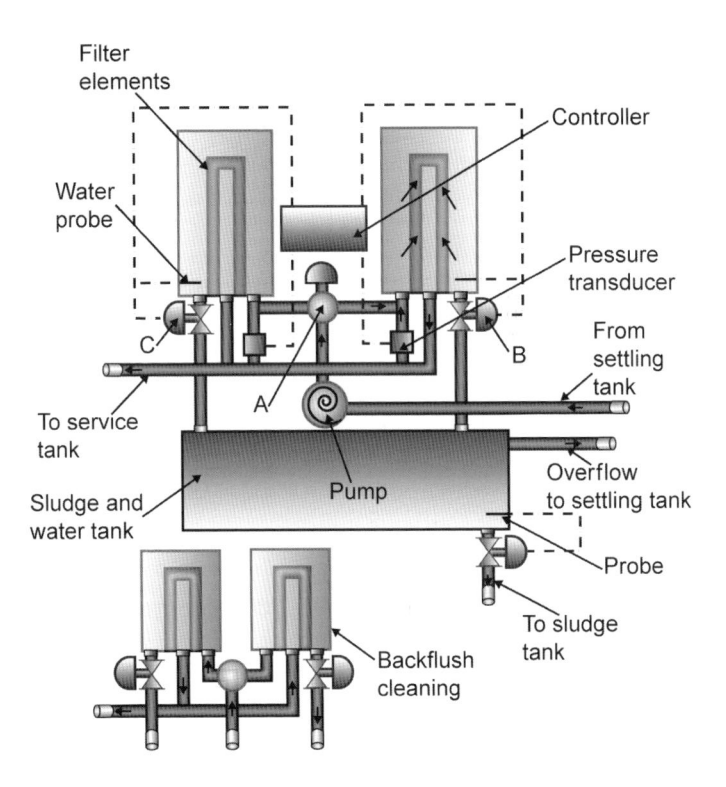

4.4 Answers – Stabilisers

1. Stabilising fins are fitted to ship's sides below the waterline to reduce rolling of the hull by wave action.

    The fins achieve this by imposing an equal and opposite motion. Retractable fins of aerofoil section use the forward velocity of the ship to create this opposing motion. As the ship rolls to starboard, the starboard fin is set by a gyroscope signal so that the leading edge of the fin is set above the axis of tilt, giving an upward thrust. The port fin is set in the opposite tilt, with its leading edge below the axis of tilt, giving a downward thrust, and vice versa when the ship rolls to port.

2. The purpose of the tail flap on stabiliser fins is to give the fin a more pronounced restoring torque action than if it were a plain surface.

3. The angle of operation of the main fin is 20 degrees at around 15 knots and 11 degrees at around 24 knots. This is to ensure reasonable fin loading.

4. The angle of operation of the tail flap is 30 degrees.

5. The hydraulic pressure used to house and extend the fin is around 70 – 80 bar.

6. The hydraulic pressure used to alter the fin angle is around 30 bar.

7. The instrument used to sense the angle of the ship and then send a signal to the fin controls is the roll sensor unit.

Sketch of fin stabiliser set up.

Engineering Pocket Book – Answers

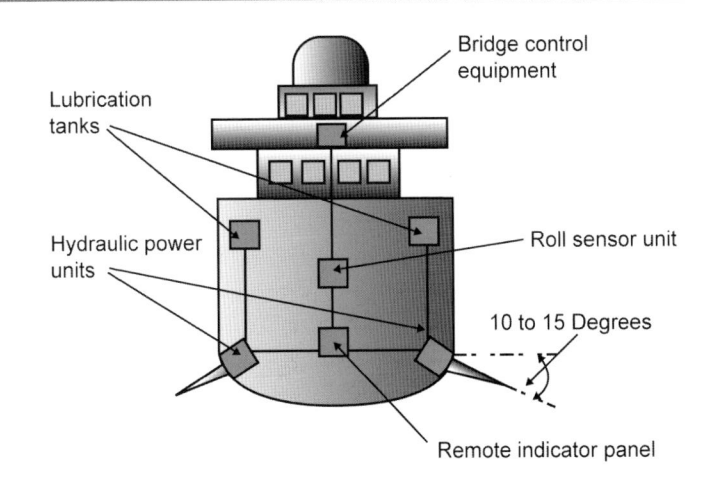

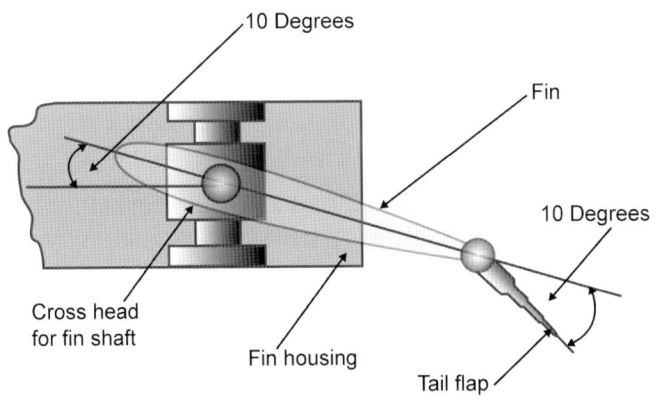

8. The panel fitted to the bridge to monitor fin position is a SOLAS panel, a positive light system to indicate if it is housed or extended. Note: this should also be backed up by visual checks.

4.5   Answers – Refrigeration

1. The four main components of a refrigeration system working on the vapour compression cycle are:

   - The compressor
   - the condenser
   - the expansion valve
   - the evaporator.

2. The function of the compressor in a refrigeration system is to raise the pressure of the vapourised refrigerant, causing its saturation temperature to rise so that it is higher than that of seawater or an air cooled condenser.

   The compressor also promotes circulation of the refrigerant by pumping it around the system.

3. The function of the condenser is to liquefy the refrigerant and sub cool it to below the saturation temperature by circulating seawater or air. Latent heat originally from the evaporator is transferred to the cooling medium.

   The liquid refrigerant still at pressure produced by the compressor passes on to the expansion valve.

4. The function of the expansion valve in a refrigeration system is to regulate the flow of refrigerant from the HP side of the system to the LP side of the system.

   The drop in pressure causes the saturation temperature of the refrigerant to fall so that it will boil at the low temperature of the evaporator.

   The expansion valve controls the flow of refrigerant to the evaporator thermostatically.

5. The function of the evaporator in the refrigeration system is to cool the air in the fridge space. It does this because the

temperature of the refrigerant entering the evaporator is lower than that of the air in the space and this causes the refrigerant to receive latent heat and evaporate.

The evaporator normally has a fan to circulate the air around it.

6. Three desirable properties of a refrigerant are:

- Low boiling point
- low condensing pressure
- high specific enthalpy of vaporisation. (This reduces the quantity of refrigerant in circulation and lower machine speeds, sizes, etc).

7. The HP cut out is fitted on the discharge side of the compressor in a refrigeration system. This will shut down the compressor in the event of an over pressure and can only be manually reset.

8. The effects of insufficient refrigerant in the system are a low reading on the LP pressure gauge and a lack of frost on the suction pipe.

5 ENGINE ROOM SYSTEMS

5.1 Answers – Bilge and Ballast System

1 & 2. Sketch of bilge system with main components itemised.

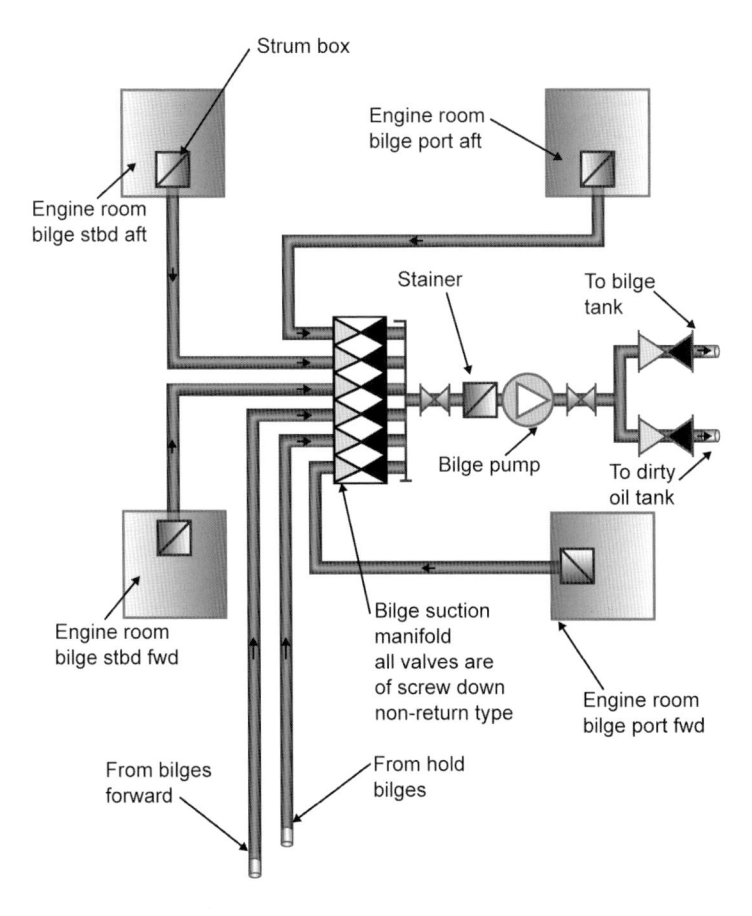

Engineering Pocket Book – Answers

3&4. Sketch of ballast system with main components itemised.

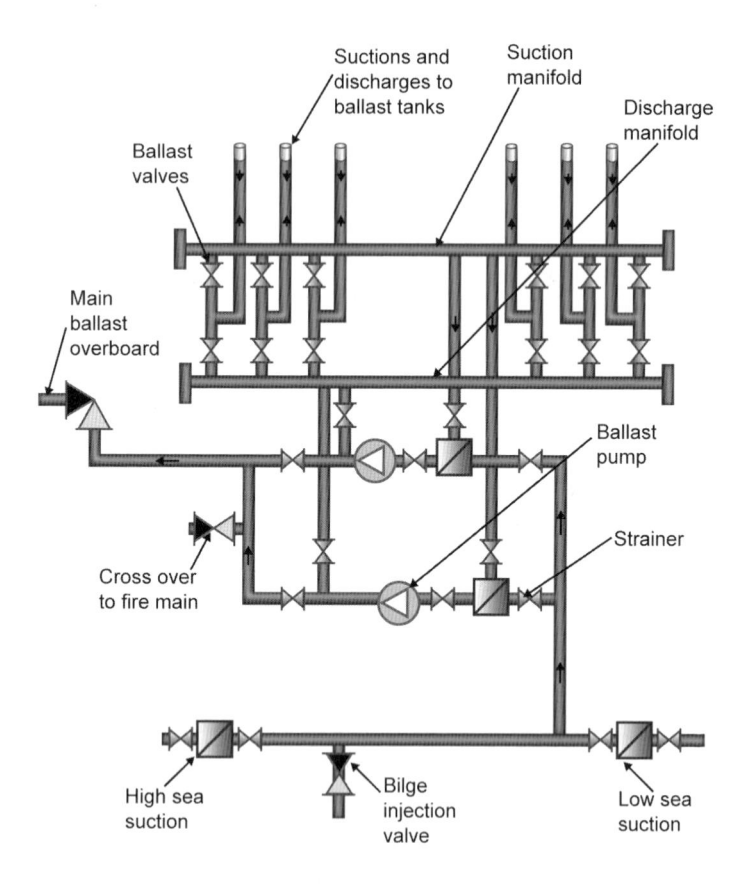

(Diagram labels: Suctions and discharges to ballast tanks; Suction manifold; Discharge manifold; Ballast valves; Main ballast overboard; Ballast pump; Strainer; Cross over to fire main; High sea suction; Bilge injection valve; Low sea suction)

5. The type of valves used in a ballast system would depend on where they were placed, but most valves would be gate valves or butterfly valves.

6. The type of valve used on a bilge system would be screw down non-return valves.

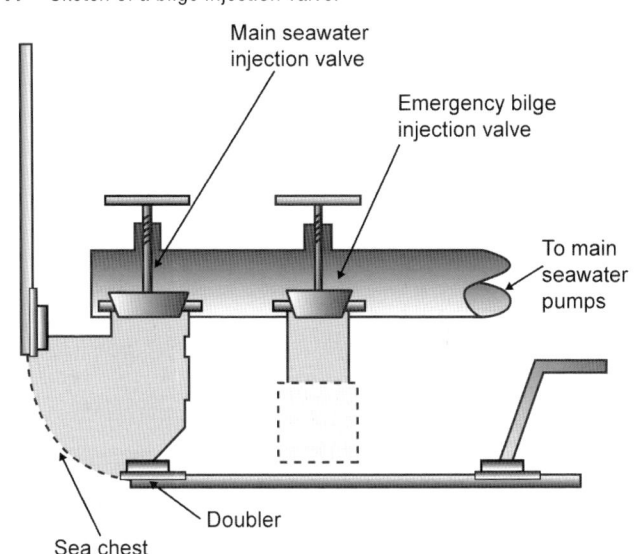

7. Sketch of a bilge injection valve.

Labels: Main seawater injection valve; Emergency bilge injection valve; To main seawater pumps; Doubler; Sea chest

8. The function of the bilge injection valve is to remove large amounts of water from the engine room in the event of serious flooding.

To do this, the bilge injection valve is opened and the main injection valve is shut, then the largest capacity pumps in the engine room will remove large quantities of water from the lowest point of the engine room.

9. The diameter of the bilge injection valve in relation to the diameter of the main injection valve should be not less than two thirds its size.

10. The function of the bilge system is to be capable of pumping from and draining any watertight compartment within the ship, except for ballast, oil and water tanks. The capacity or size of pumps in the system depend on the size, type and service of the ship.

11. The function of the ballast system is to pump to and empty all spaces in the ship in which ballast water can be supplied. The suction pipes for the ballast system will be completely separated from the bilge system. The ballast system can also be used for other purposes in emergencies, such as supplying water to the fire main.

5.2 Answers – Fresh Water and Seawater Cooling System

1. The advantages of a central cooling system are;

   - Less maintenance, due to fresh water system having cleaned treated water
   - fewer salt water pumps with attendant corrosion and fowling problems
   - simplified and easier cleaning of coolers
   - higher water speeds possible with a fresh water system, resulting in reduced pipe dimensions and installation costs
   - the number of valves made of expensive material is greatly reduced, also cheaper materials can be used throughout the system
   - constant level of temperature is maintained, irrelevant of seawater temperature, also no cold startings, reduced cylinder liner wear, etc.

2. Four checking and treatment procedures used to reduce the possibility of damage to freshwater systems are:

   - Cleaning of cooling water chambers and tanks
   - analysing the applied untreated freshwater
   - treating the applied water
   - checking the water in the cooling system.

3. Corrosion may occur in freshwater systems as a result of makeup water from the evaporators, as this water is very soft

Engineering Pocket Book – Answers

and can absorb large quantities of CO_2 which reduces the PH value of the water.

The water may also contain small amounts of chlorides.

4. The purpose of an expansion tank in a freshwater system is to allow for venting and pressurisation of the system and it also makes up for any losses.

5. Sketch of fresh water cooling system for a diesel engine that does not use a central cooling system:

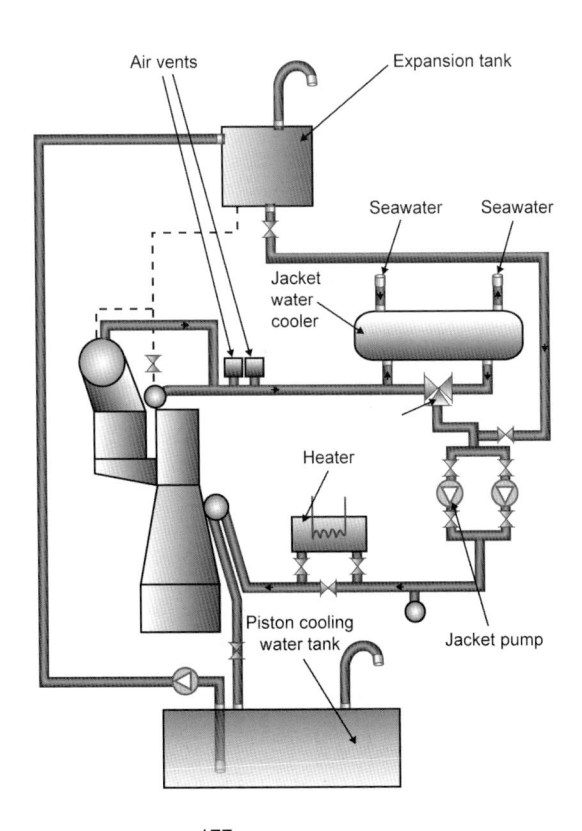

Engineering Pocket Book – Answers

6. The upper sea suction valve is used while in port to prevent any mud or sand entering the cooling system. Its also used while sailing in shallow waters.

7. The lower sea suction valve is used when sailing in deep water to avoid air entering the cooling system while the ship is rolling or pitching.

8. Steam or air may be supplied to the seachest to remove any blockage such as seaweed, sludge, etc.

9. Sketch of a system that makes use of auxiliary engines to heat main engines while shut down:

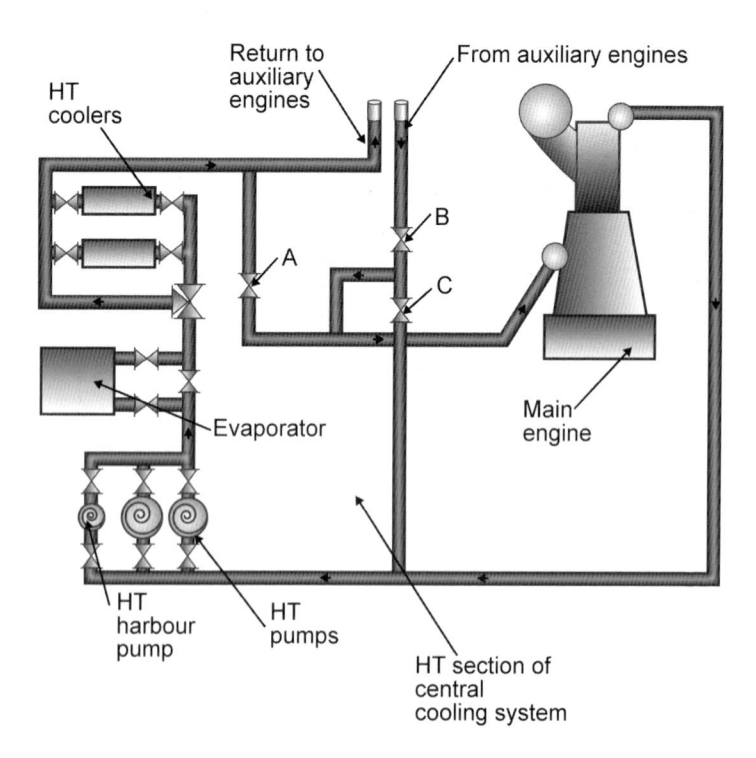

10. The advantages of the above system is that a heater is not required for keeping the main engines warm, the engines are always ready for use.

5.3 Answers – Fuel Oil System

1. Sketch of boiler fuel oil system:

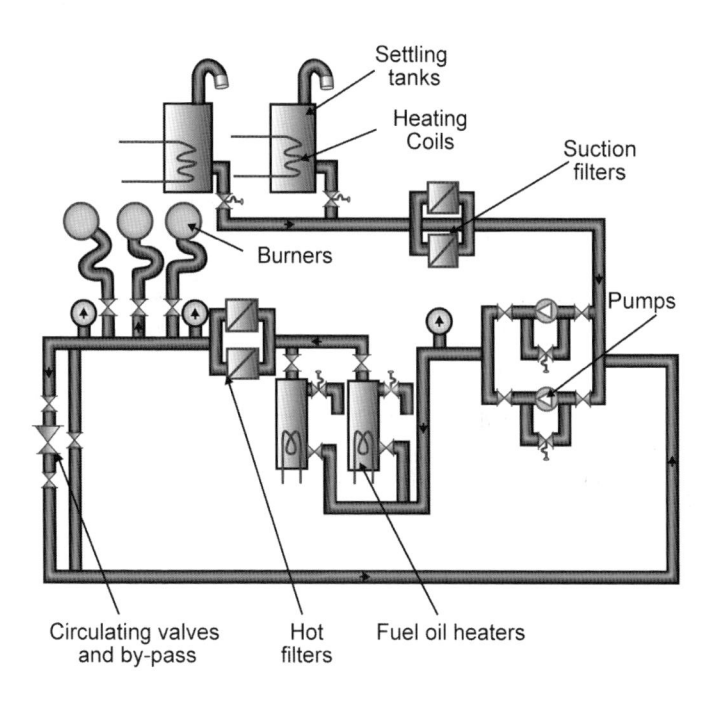

Settling tanks, Heating Coils, Suction filters, Burners, Pumps, Circulating valves and by-pass, Hot filters, Fuel oil heaters

Engineering Pocket Book – Answers

2. Sketch of diesel engine fuel system:

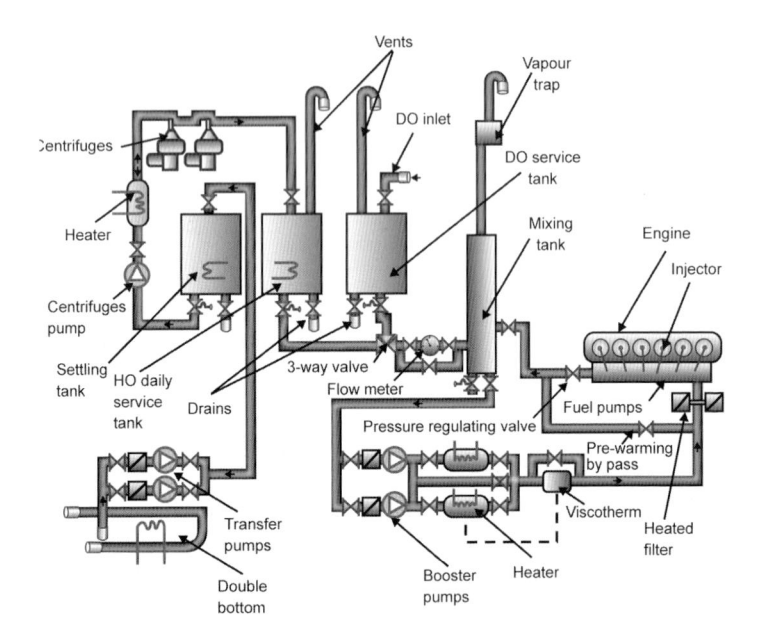

From the bunker tanks, fuel is transferred by the transfer pump to the settling tank, from the settling tank, fuel oil is purified to the service tank.

From the service tank, the fuel oil is pumped through a pressurised fuel system to the engine.

The fuel oil firstly passes through a set of cold filters to a set of fuel oil booster pumps, raising the fuel oil pressure to around 12 – 15 bar, delivering the fuel through a set of heaters and viscotherm, a set of fine filters then to the fuel rail and to the engine fuel pumps where the pressure is raised to around 250 – 300 bar for atomisation by the fuel injector.

Engineering Pocket Book – Answers

The heater in the system reduces the fuel oil viscosity in the system for efficient combustion. The temperature required will depend on the fuel oil quality which will vary, however the temperature should not exceed 150°C. The fine filter in the system is a stainless steel mesh to filter out particles larger than 50 microns, or less for smaller engines. Filters should be cleaned regularly.

3. Sketch of a settling tank:

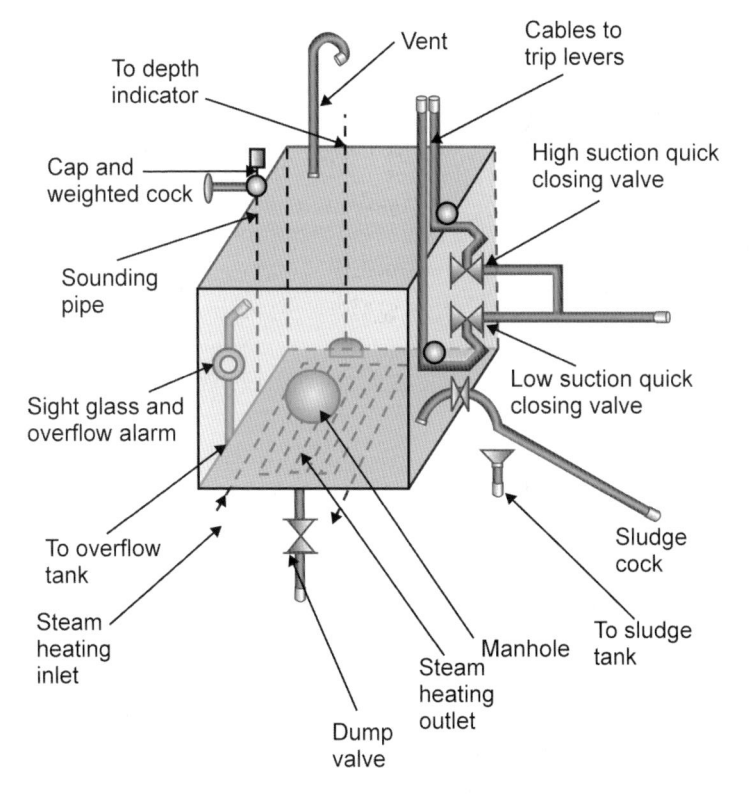

Labels: Vent; Cables to trip levers; To depth indicator; Cap and weighted cock; High suction quick closing valve; Sounding pipe; Low suction quick closing valve; Sight glass and overflow alarm; To overflow tank; Sludge cock; Steam heating inlet; Manhole; To sludge tank; Dump valve; Steam heating outlet

4. Settling tanks use gravitation for the purification of fuel oil. When the oil is allowed to stand undisturbed in the tank, particles and liquids of higher density than oil will gravitate to the bottom of the tank where they are discharged every so often by a manually operated spring-loaded sludge cock. This process can be speeded up by applying heat to the tank's contents.

Steam heating coils are provided for this purpose, but care must be taken that the oil is not heated to too high a temperature. These tanks are normally lagged to keep the heat in.

5. The temperature of fuel oil is raised before burning to reduce the high viscosity of the fuel to a value at which correct atomisation can take place by the fuel injectors.

   This will allow correct mixing and burning for efficient combustion. Viscosity of fuel oil can be reduced by passing it through a heater.

6. Correct temperature is maintained by the viscotherm which automatically regulates the temperature of the heater to a constant temperature to maintain controlled viscosity.

7. The two emergency valves fitted in a boiler fuel system are:

   - A quick closing valve on the settling tank which is operated from outside the engine room
   - safety valves or cocks on each individual burner – this valve will operate in the event of high steam pressure, low water level, high water level, flame failure.

   Emergency valves on burners have manual shut off for fuel, and automatic shut off in event of boiler fault.

8. The purpose of the recirculating valve fitted in a boiler fuel system is to enable the fuel oil to be brought up to the desired temperature quickly for combustion.

9. Safety fittings on a boiler fuel system are:

- Relief valves on heaters
- spring-loaded relief valves on pumps will spill back to suction side of pump
- all pipework is lagged
- save-alls under pumps/heaters
- quick closing valves on settling tanks
- manual quick shut off valves on burner
- high fuel oil temperature alarm
- low fuel oil temperature alarm.

10. The boiler fuel oil system can be changed to manual control by means of by-passing the automatic control valve and controlling the fuel oil supply pressure by means of a hand jacking valve on a spring-loaded relief valve governing the discharge pressure from the fuel oil service pumps.

11. The density of fuel oil burned in a diesel engine is important because some fuels of different densities are not compatible and formation of heavy sludges can occur in tanks.

12. The viscosity is important because it is required to be able to calculate temperatures at which fuel is treated and injected into the engine for correct combustion.

13. The flash point in a fuel system is important for safety reasons. The flash point of any fuels onboard ship must not be less than 60°C.

14. Solid contaminants that might be found in fuels are rust, sand, dirt or refinery catalysts which are all abrasive materials and cause wear.

15. Liquid contaminants found in a fuel oil system may be salt or fresh water.

Engineering Pocket Book – Answers

16. Two centrifuges are fitted to a diesel engine fuel system in series because the first one is used as a purifier that removes water and sludge from the fuel oil and the second centrifuge is used as a clarifier, which removes solids from the fuel oil.

17. Various safety devices in a fuel system for a diesel engine are:

    - Quick closing valves on settling/service tanks
    - relief valves on 2 pumps/heaters
    - quick closing valve on mixing/vent tank
    - pipes lagged/save-alls under pumps and heaters
    - low fuel oil pressure alarm
    - high fuel oil pressure alarm
    - low fuel oil temperature alarm
    - high fuel oil temperature alarm
    - emergency remote stops for pumps
    - high pressure pipes between fuel injection pump and injector are double skinned.

18. Viscosity control for the temperature of a fuel oil heater is considered more superior because it is possible that oils stored in the ship's tanks or even one tank could have come from different sources and have different properties, therefore viscosity control is considered more superior.

5.4 Answers – Engine Room Systems (General)

1. Two reasons why fuel pipes should be clipped and supported are to prevent stress and fractures from vibration.

2. The reason for circulating fuel in large two-stroke engines is to ensure the system is fully primed and at temperature.

3. Oil that has been recirculated in a diesel engine fuel system normally returns to a buffer/vent/mixing tank.

4. When burning heavy fuel oil in a diesel engine, it is necessary to reduce the high viscosity of the fuel oil to a value at which correct atomisation and combustion can take place.

This will allow for correct mixing and burning for efficient combustion.

5. The recommended standard treatment of residual fuel to be used in large diesel engines is to first allow the fuel to settle in a settling tank where any water/sludge can be drained off. It is then purified and clarified to the service tank, the purifier further removing any water or sludge and the clarifier removing any solids.

Before reaching the engine from the service tank, it will pass through a fine filter which will remove any particles larger than 50 microns, or even less for smaller engines.

6. It is advisable to store ship's bunkers from different ports in separate bunker tanks as they may not be compatible. If this precaution is neglected, there is a risk of heavy sludge formations occurring due to incompatibility and not mixing successfully.

7. The thermostatic control on a boiler fuel oil heater controls the amount of steam heating to the heater.

8. Boiler fuel oil systems use gas oil to flash up from cold and this is achieved by supplying air to the burner instead of steam until sufficient steam pressure is achieved, then steam can be supplied to the heavy fuel oil heaters. The heavy fuel oil is circulated until the correct temperature is achieved, then steam is supplied to the burner instead of air, plus heavy fuel oil.

9. The function of the dumping valve fitted to a fuel oil settling tank is to dump the oil from an elevated settling tank to a lower bunker double bottom tank in the event of a fire in the engine room or close to the tank.

Engineering Pocket Book – Answers

10. The function of the sludge valve or sludge cock fitted to a settling tank is to drain water or sludge at the bottom of the tank to any oily bilge or sludge tank.

    The cocks are of a self-closing type.

11. The function of the steam trap from heating coils is to ensure maximum utilisation of the steam from heating coils. From the steam trap, condensate passes to an observation tank for oil leakage detection.

12 & 13. Sketch of an automatic domestic water supply system.

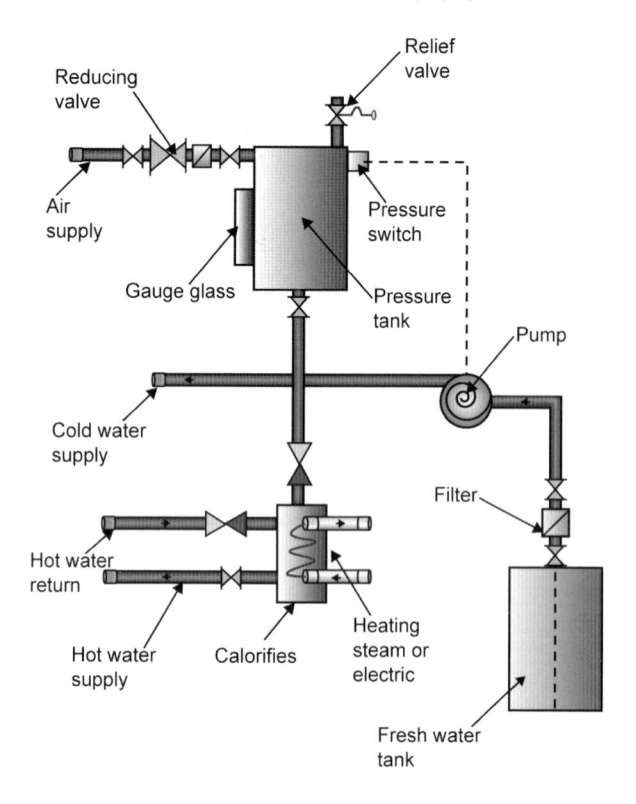

14. The fresh water pumps take suction from the tank that is in use, pump the water via a neutraliser, which makes the water slightly alkaline and improves taste, and a hypochlorinator to sterilise the water, to the pneupress tank, which comprises of water and an air space at the top and is pressurised to around 4 bar with compressed air.

As the water level drops, the pressure drops and when the pressure drops to around 3 bar, a pressure switch is activated which starts the fresh water pump. The water level in the pneupress tank will then rise, causing the pressure to build up, and once it reaches 4 bar another pressure switch is activated that shuts down the fresh water pump.

From the pneupress tank, the water passes through a sand bed and carbon bed filter to absorb any excess chlorine in the water. It then branches off to the cold supply or through a calorifier for the hot water supply.

Note that all these filters can be backflushed by arranging the cocks on them for cleaning.

15. In domestic fresh water systems, carbonates of calcium and magnesium are used to produce water that is slightly alkaline and to improve its taste.

16. Chlorine is used in fresh water systems to sterilise the water, the amount used is a solution of 100000 : 1.

17. Sketch of a central priming system:

The water ring exhausters maintain the vacuum in the tank with preset limits. Opening of the SDNR valves or cocks for a pump will cause priming to take place. To prevent water entering the vacuum tank after priming has taken place, air release valves are fitted which automatically close.

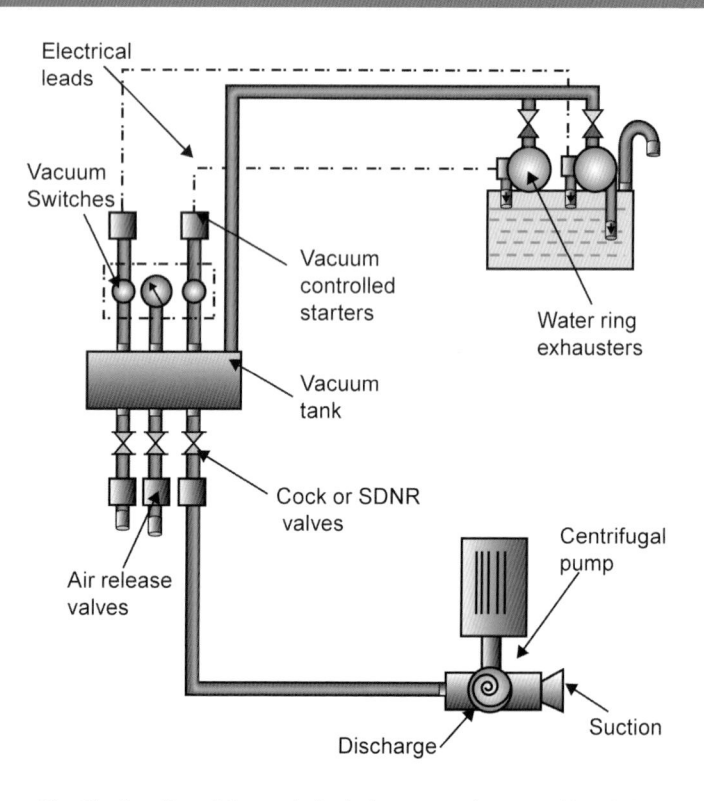

18. The function of the central priming system is to provide priming for all centrifugal pumps located in the engine room.

19. The advantages of a central priming system are:

- Saving in power as each pump does not require its own exhauster
- reduced capital cost
- simplified maintenance
- automatic.

Engineering Pocket Book – Answers

20. The water is prevented from entering the vacuum tank after priming by means of air release valves that close automatically.

21. British Standard symbols:

- Screw down valve
- non-return valve
- safety valve
- butterfly valve
- double seated change-over valve.

22. British Standard symbols:

- Suction box
- filter or Strainer
- steam trap
- separator.

Engineering Pocket Book – Answers

23. British Standard symbols:

- Funnel

- open scupper with closing device

- pipe flow indicator

- pressure gauge

- thermometer.

24. Bilge and ballast systems are interconnected so that each can perform the other's function in an emergency, ie a ballast pump could be used to pump out a flooded engine room.

25. They are connected by means of a cross–over valve.

6 SHIP CONSTRUCTION

6.1 Answers – Ship Construction

1. Longitudinal sketch of a vessel, identifying the principal parts:

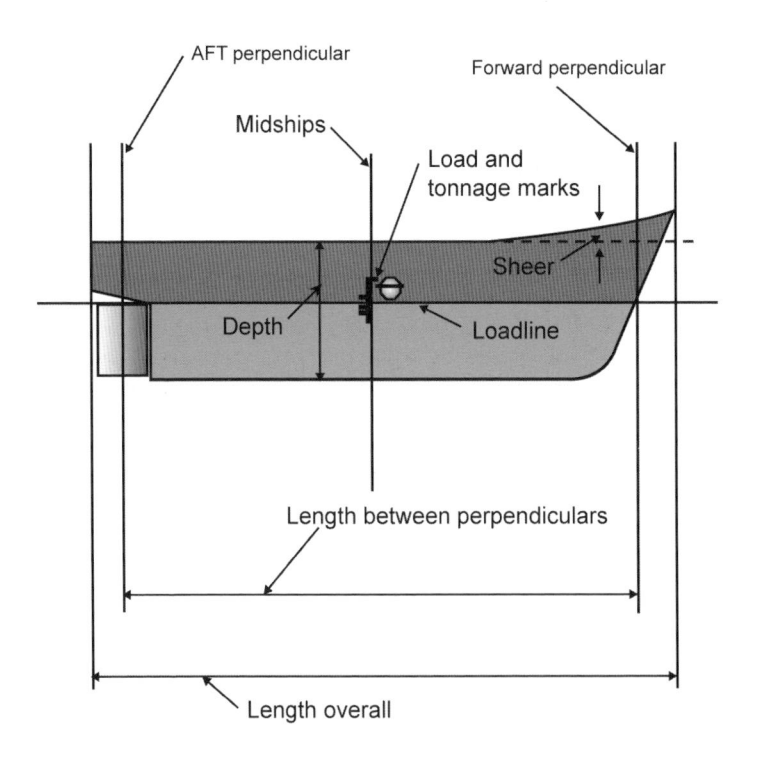

Engineering Pocket Book – Answers

2. Transverse sketch of a vessel, identifying the principal parts.

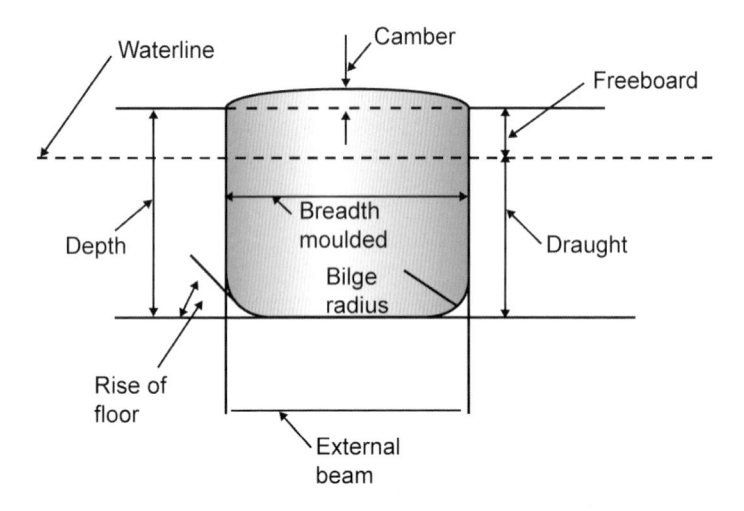

[Diagram showing a transverse cross-section of a vessel with the following labels: Waterline, Camber, Freeboard, Depth, Breadth moulded, Draught, Bilge radius, Rise of floor, External beam]

3. The loadline disc is positioned midships, placed directly below the deck line. The distance from the upper edge of the deck line to the centre of the disc is the statutory summer freeboard, this is also known as the loadline.

4. Panting is used to describe the in and out movement of the shell plating caused by fluctuations in water pressure caused by waves as they pass along the ship.

5. The difference between a strake and a stringer is a strake is a line of vertical plating extending forward and aft and a stringer is a line of horizontal plating extending forward and aft.

Engineering Pocket Book – Answers

6. The purpose floors serve are as transverse vertical stiffeners which strengthen the ship's bottom plating and which may be enclosed to form double bottoms.

7. With regard to pipe-work, the advantage of a duct keel is that pipe-work can be run through it and it also allows the pipe-work to be accessible when cargo is loaded.

8. Two purposes of a double bottom are to provide a safety margin if so the vessel was damaged on the shell bottom due to grounding, only the double bottom would be flooded. It can also be used to carry fuel oil, fresh-water and to provide ballast capacity.

9. The area of the ship where the number of floors increase maybe where heavy loads are supported or local stresses occur, ie holds of bulkers, engine rooms, etc.

10. Bilge keels are fitted to help damp the rolling motion of the vessel. Other relatively minor advantages of the bilge keels are protection for the bilge on grounding and increased longitudinal strength at the bilge.

11. The types of steel used in tanker construction are mainly mild steels throughout, but higher tensile steels may also be introduced in more highly stressed regions of the larger vessels.

12. In the construction of tankers, Lloyd's Register generally requires full longitudinal framing in tankers of all lengths.

13. In tankers, cross ties are fitted in wing tanks to connect the vertical webs at the ship's side and longitudinal bulkhead. The cross ties are designed to stiffen the tank side bounding bulkhead structure against transverse distortion under liquid pressure.

Sketch of cross ties.

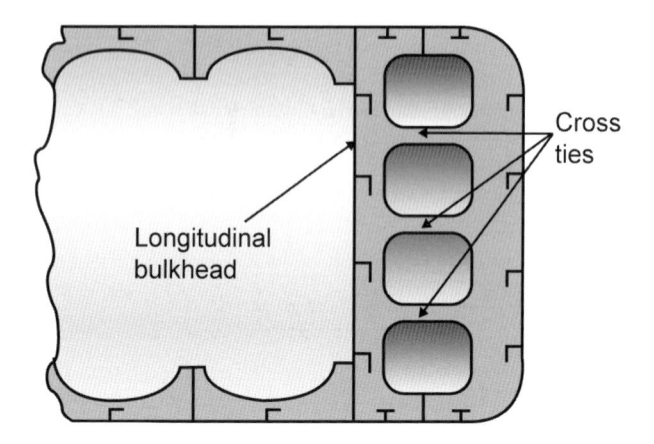

[Diagram showing cross-section with longitudinal bulkhead and cross ties labeled]

14. The spaces that may be accepted in lieu of cofferdams in tankers are pumprooms and ballast tanks.

15. A secondary barrier may be required as a lining between the cargo container and the ship's hull in a gas carrier because a leak from the cargo could damage the ship's hull due to low temperatures and also as a safety measure for personnel.

16. Membrane tanks in a gas carrier are non-supporting tanks consisting of a thin layer (membrane) supported by insulation by the adjacent hull structure. The membrane is designed in such a way that thermal and other expansion and contraction is compensated for without due stressing of the membrane. Membrane tanks are primarily used for LNG cargos.

Sketch of a membrane tank.

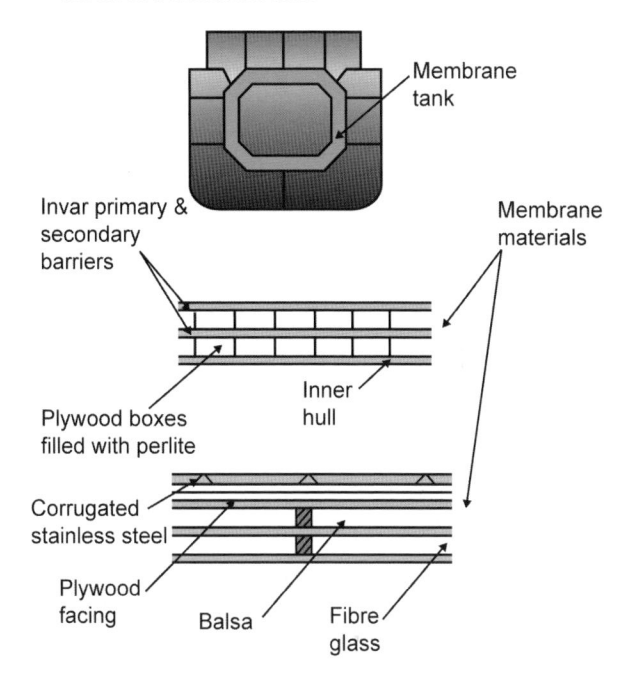

17. The materials used in construction of liquefied gas tanks are Invar and aluminium alloy.

18. If a ship were steam powered, the 'boil off' vapour from an LNG ship's tank could be used as fuel for the boiler.

19. The collision bulkhead on a ship would be found no more than 1/20th of the ship's length away from the forward end of the ship.

20. The safety notice that should be displayed at each MacGregor type hatch is 'Do not remove locking pins until check wire is fast and all persons are clear.'

Engineering Pocket Book – Answers

21. Sketch of a watertight door.

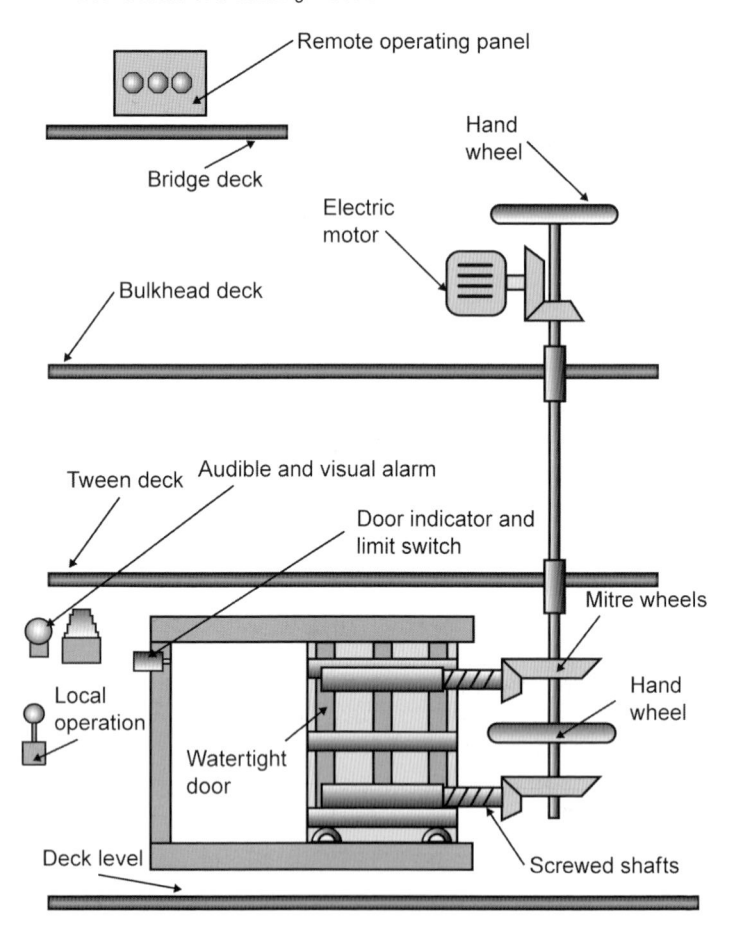

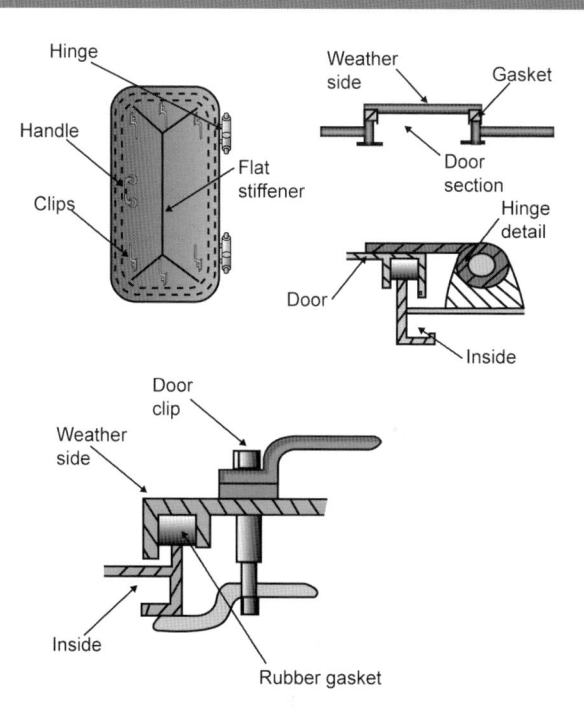

22. The arrangement made to allow personnel to pass through a watertight bulkhead are watertight doors, which will be a sliding type of door, operated hydraulically or by electric motor. These can be operated locally by the door or either remotely at an emergency station or from the bridge.

23. Three Classification Societies could include:

- Lloyds Registrar
- American Bureau of Shipping
- Det Norske Veritas
- Bureav Veritas
- Germanscher Lloyd
- RINA

7 STEERING GEAR

7.1 Answers – Steering Gear

1. The three basic requirements of steering gear are

   - To be continuously available, move the rudder rapidly to any position of degrees in response to the order from the bridge during manoeuvring and hold it in the required position
   - have arrangements for relieving abnormal stress and returning it to its required position
   - maintain the ship on course regardless of wind and waves.

2. Jumping stops are fitted to prevent the lifting of the rudder and stock in heavy weather.

3. In ram type hydraulic steering gear, load is relieved on the cylinder glands by the crosshead slippers bolted to the side of the rams, which slide on a machined surface of the guide beam.

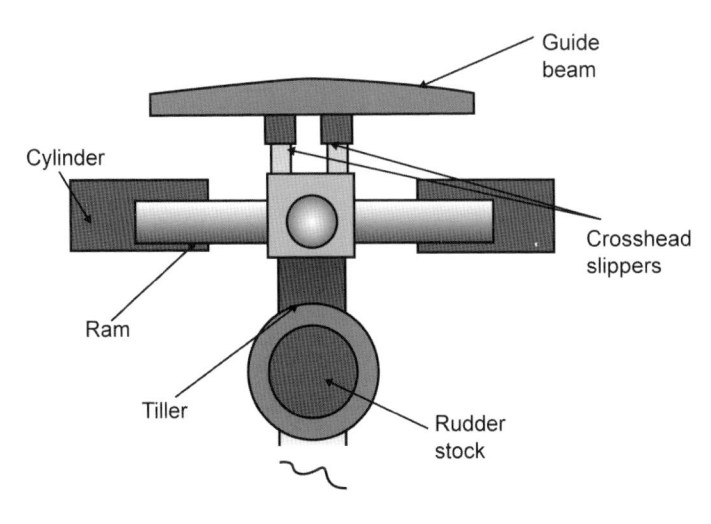

Engineering Pocket Book – Answers

4. The cylinders are braced together by the guide beam because hydraulic pressure tends to push them apart. The cylinders also have feet that are bolted to stools on which the gear is mounted. In four cylinder sets, adjacent cylinders are cross braced by heavy brackets.

5. In an emergency in hydraulic steering, the rudder can be locked in position by closing the supply valves to each ram.

6. Hydraulic steering gear is protected against overloading by heavy seas acting on the rudder by relief valves which operate around 10% above normal working pressures, oil being passed to one side of the system to the other.

The hunting gear will correct the rudder movement putting the steering gear pump on 'stroke.'

7. In hydraulic steering gears, attention should be paid to high pressure oil pipes because, in some cases, it has been known for the pipes to fracture at necks, especially after heavy weather.

8. The rudder angle permitted by rotary vane steering gear with two moving and two fixed blades is 75 degrees to port and 75 degrees to stbd, and in three moving and three fixed blades is 35 degrees port and 35 degrees stbd.

9. In a rotary vane steering unit, the rotor is attached to the rudder by means of taper and key. The rudder stock is tappered in way of the rotor which is keyed to it.

10. The vanes are sealed in a rotary vane steering unit by steel strips backed by synthetic rubber laid in slots.

11. Rotation of the stator in a rotary vane unit is prevented by two anchor bolts held in fixed anchor brackets with rubber shock absorbing sleeves.

Engineering Pocket Book – Answers

12. In the Hele-Shaw pump, there are usually between seven and nine radial cylinders.

13. Sketch of a Hele-Shaw pump.

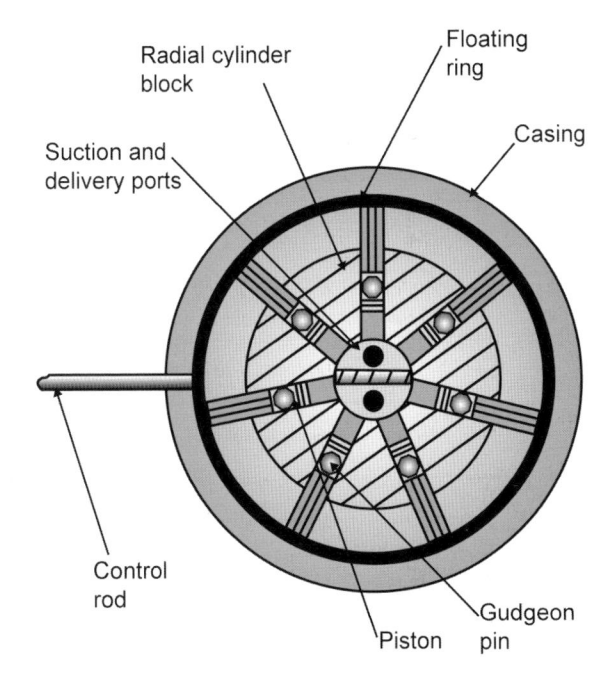

Labels: Radial cylinder block, Floating ring, Casing, Suction and delivery ports, Control rod, Piston, Gudgeon pin

The pump consists of seven or nine radial cylinders which are rotated at constant speed in one direction. The radial cylinder block rotates on a fixed steel central piece having two ports opposite one another and in line with the bottom of the rotating cylinders.

In each cylinder is a piston with a gudgeon pin with bronze slipppers on the end. The slippers revolve with the cylinder block in grooves machined in the floating ring.

Movement of the floating ring displaces the circular path of rotation of the pistons from that of the cylinder block, and produces a pumping action. When the rod is in the mid postion and the centres of rotation of the pistons and block coincide, there is no pumping.

14. The flow of oil in a VSG (varible speed gear) pump is varied.

The VSG pump consists of a cylinder block with axial cylinders, piston stroke and oil flow are varied by the angular movement of the swash plate.

A simplified sketch of a VSG pump.

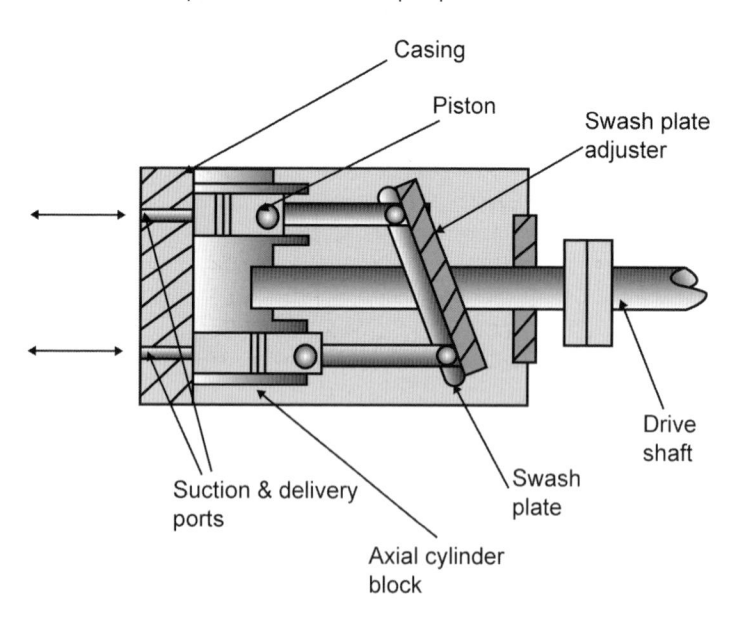

15. The idle pump is prevented from running in the reverse direction when two pumping units are fitted by means of non-reverse locking gear. This is fitted between the pump and the motor in the flexible coupling and consists of steel pawls on the motor coupling rim. These pawls, when the pump and motor are running, are thrown out towards the rim by centrifugal force. With the pump stopped, the pawls return to their normal position and engage the teeth of a fixed ratchet secured to the pump base.

16. The mechanism that restores the rudder postion after it has been displaced by heavy seas is the hunting gear.

If the rudder is displaced by heavy seas, the hunting gear is moved by the rudder stock, this will put the pump on stroke and the rudder will be restored to its previous position.

17. The three types of control systems for electro-hydraulic steering gears are:

- Non-follow up systems – in these systems, the gear will run and the rudder will continue to turn while the helm is from its central position, rudder movement will only cease when the helm is centred once again
- follow up systems – with these systems, movement of the rudder follows movement of the helm. For example, if the helm is moved to a desired position, the rudder will turn until this position is reached and cease, the helm remaining offset to its central position
- automatic systems – with these systems, the steering control circuits are controlled by signals received by the master compass so that the ship is automatically held on its selected course

18. The working fluid in a telemotor control system is mineral oil of low viscosity and pour point as this gives protection against rusting. As an alternative to mineral oil, a mixture of glycol (glycerine) and water can been used.

Engineering Pocket Book – Answers

19. The precaution taken when charging hydraulic steering gears with oil is that all the air is purged from the system.

20. Control is effected on an electro-hydraulic steering gear with fixed capacity pumps by means of directional and proportional valves.

21. According to international regulations, the steering gear must be tested within 12 hours before each departure (except for a ship on short voyages where tests are carried out weekly). Also, emergency drills must be carried out every 3 months and within 24 hours before a vessel enters US waters.

22. Before testing the steering gear, the deck officer or department must be informed.

23. Before allowing the test, they will ensure that there are no obstacles in the way of the rudder.

24. Steering gear tests:

- Deck department is informed
- communications between bridge and steering flat checked
- number 1 pump started
- steering gear moved from midships to 35 port and back to midships, then to 35 stbd, rudder angle been confirmed on movement
- number 1 pump is stopped, which will set off steering gear failure alarm, number two pump should cut in, same test as before is carried out on number 2 pump, confirming rudder angles
- both pumps are then started and tested, confirming rudder angles
- while carrying out these tests, all pipework, linkages and greasing of linkages should be checked
- oil level in supply tank must be checked and be at least 75% full

25. Power to the steering motors must be by two duplicated, widely separated supplies from the main switchboard, one of which may come from the emergency switchboard.

26. The electrical protection given to steering gear circuits is short circuit protection only. The bridge will have failure alarms fitted, with manual or auto start up of the power units.

8 SHAFTING

8.1 Answers – Shafting

1. Propulsion shaft bearings are divided into two groups which are those inside the watertight integrity of the ship and those that are outside the hull watertight boundary, ie shaft and stern tube bearings.

2. Shaft bearing loads do not vary with RPM but are essentially constant at all speeds.

3. Reliability is particularly important in shaft bearings because there is no redundancy for bearings and a single failure may incapacitate the propulsion system.

4. Shaft support bearings are normally lubricated by means of self-lubrication where rings or discs are arranged in such a manner that lubrication is effected by rotation of the shaft.

5. The advantages of using roller bearings for shaft support are that they are lighter in weight and have a lower coefficient of friction but can only be used on smaller shafts. In general, the advantages are not sufficient to offset the higher reliability and lower maintenance costs of the Babbitt lined type.

6. Roller bearings are not used to support larger types of shaft because they are not as reliable as Babbitt lined types.

7. Sketch of self-aligning propulsion shaft bearing:

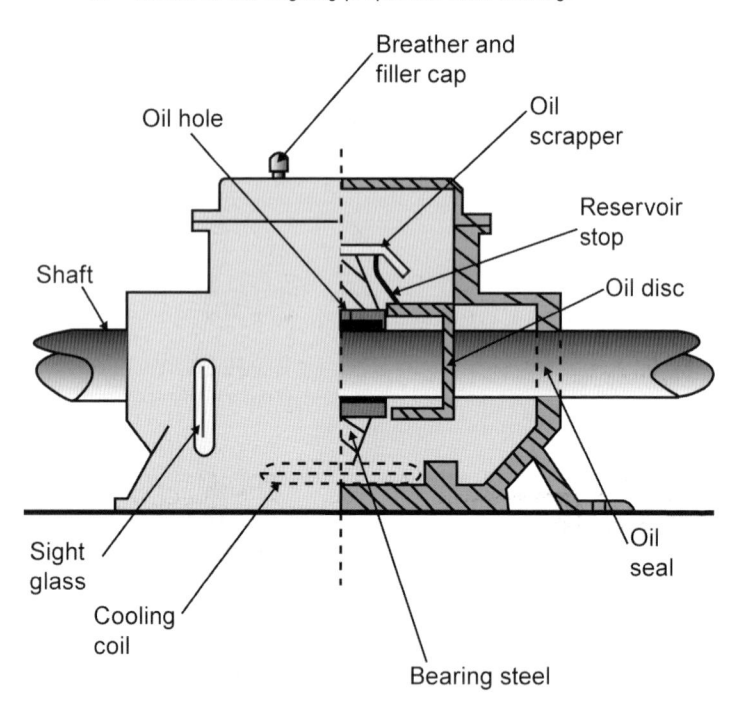

(Diagram labels: Breather and filler cap, Oil hole, Oil scrapper, Reservoir stop, Oil disc, Shaft, Sight glass, Cooling coil, Bearing steel, Oil seal)

8. A propulsion shaft may sag or hog while in use due to the state of loading, state of draught or the effect of waves.

Engineering Pocket Book – Answers

9 & 10. Sketch of an oil lubricated stern tube and stern tube lubrication system (Transmission shafting system).

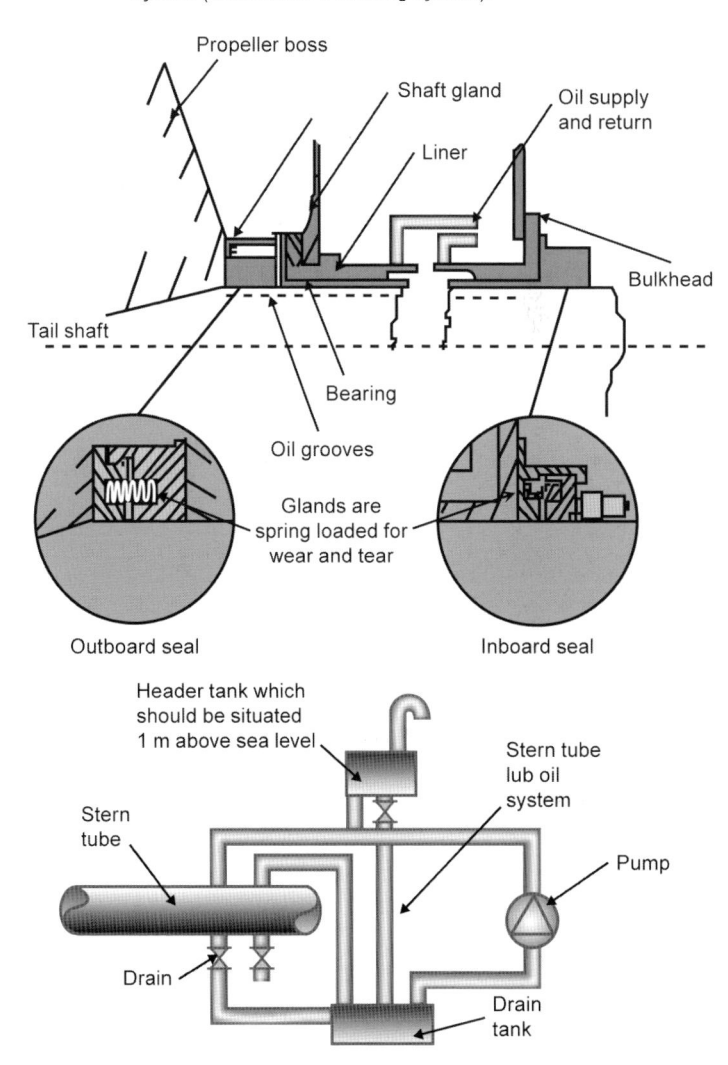

Engineering Pocket Book – Answers

11. The holding down bolts of a thrust block are relieved of sheer by the wedges at the base of the block, which have a slow taper of 20 mm/m and act to relieve sheer.

12. Sketch of a ship side valve.

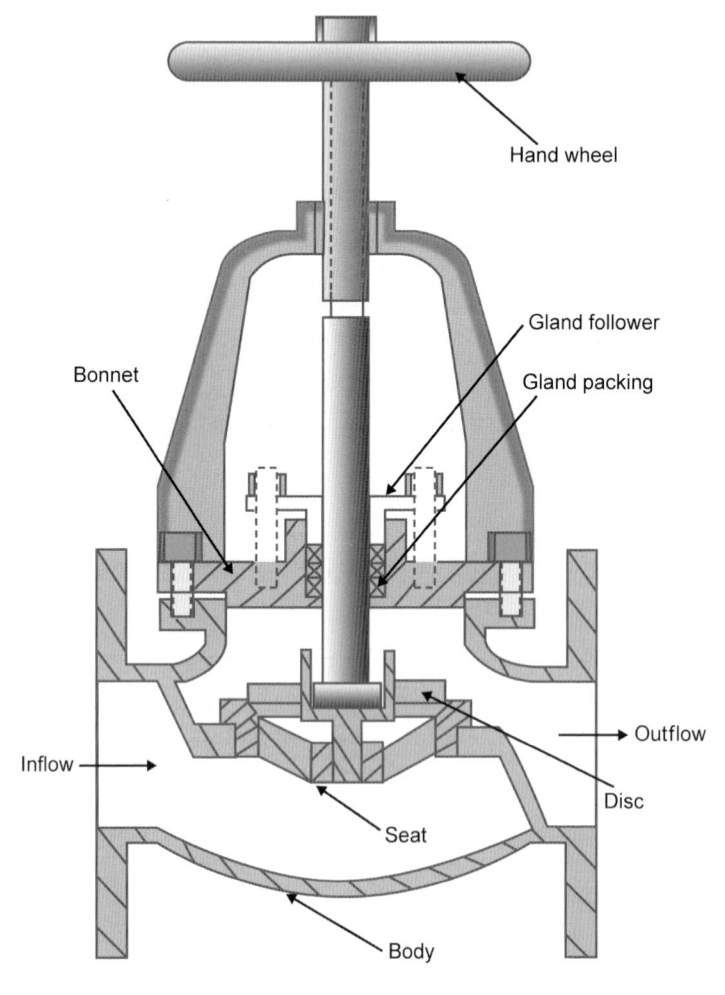

- Hand wheel
- Gland follower
- Bonnet
- Gland packing
- Inflow
- Outflow
- Seat
- Disc
- Body

This sketch shows a ship side valve, which must be a non-return type. Generally, the valves will be of the type where the closing device consists of a circular disc with some arrangement of wings, or a centrally located guide rod on the base of the disc.

The wings or the guide rod guide the disc in its up and down motion in and out of its seat. The disc is sometimes referred to as the valve lid or the clack, and is the closing device that prevents the return flow through the valve.

These valves are generally made of non-corrosive materials.

13. Screw down non-return valves are used as ship side valves because they prevent flooding of the vessel if a piping system should fracture or break with the discharge submerged, they also allow for isolation.

14. The advantages of hydraulic-powered actuators are can that they operate large valves that are beyond manual operation and can operate valves remotely, ie from the engine control or cargo control room.

15. Damage to valves with powered actuators is prevented by means of limit switches that cover the movement of the actuator/valve.

16. Care and maintenance of ship side valves:

- Care and maintenance of valves can mean valve seats being cleaned due to dirt or scale build up or worse, where the seat may require regrinding. Any of these conditions would make the valve pass/leak
- leakage at the stuffing box/gland can be dealt with by nipping up the gland nuts, but not too tight that the valve stem sticks, or even by repacking the gland. If this fails, it could mean that the stem is marked or bent

- sticking of the valve stem could mean the packing is too tight in the gland, or it could be caused by the valve being shut tightly while the valve is hot and, when it has cooled, contracted causing the disc to bond tightly with the seat. The same could happen if the valve is jammed open while cold. Subsequent heating can cause expansion which binds the valve open and a wrench should open or close the valve in this case
- when opening a valve, it is good practice to shut it in slightly, ie ¼ turn
- a serious condition that can be caused by rough handling is damage to the threads of the stem. Sometimes, this condition can be fixed by dressing the threads with a file or, if severe, the stem will probably have to be replaced

9 ELECTRICAL

9.1 Answers – Switchboards, Circuit breakers and Switchboard Instrumentation

1. The duty of a switchboard is to distribute the generated electricity to where it is required.

2. The purpose of a circuit breaker is to interrupt fault current as quickly as possible and so keep damage to other pieces of equipment to a minimum.

3. The duties of the springs in a circuit breaker are to open the contacts in the tripping operation.

4. The current rating used with MCCBs is 30 – 1500A.

    The current rating used with MCBs is 5 – 100A.

    The current rating used with a generator breaker or other large breakers is 600 – 6000A.

5. The maintenance carried out on MCBs is none, as no maintenance is possible on them.

6. An ammeter measures current.

7. A wattmeter measures power.

8. The voltage in the secondary winding of a voltage transformer is 220V (this may vary from ship to ship).

9. When voltages are equal in frequency, the synchroscope pointer lies at the noon position.

10. The maximum time a synchroscope should be on line for is 20 minutes, as it is not continuously rated.

Engineering Pocket Book – Answers

9.2 Answers – Parallel Operation of Generators

1. The circuit breaker must be closed when paralleling when the voltages and frequencies are the same and the pointer on the synchroscope is at the 11 o' clock position.

2. When using a synchroscope, the pointer should be travelling in a clockwise direction at no more than 1 rev/5 seconds. This ensures the incoming machine is slightly fast and it will immediately assume load.

3. When using synchronising lamps, the correct moment to close the breaker is when the key lamp is dark (top light when in triangular formation) and the other two lamps are equally bright.

4. After successful synchronisation, the generator load should be shared equally.

5. The generator setting is varied to balance generator load.

6. To adjust the power factor, the excitation of the alternator is adjusted.

9.3 Answers – Motor Starters

1. The most common type of starter at sea is the direct on line starter (DOL).

2. The starting current compared to full load current is 5 – 8 times greater.

3. When connected in star, the voltage as a percentage of the full load voltage is 58%.

4. The most common reduced voltage starter is the star–delta starter.

Engineering Pocket Book – Answers

5. With a manual change-over in a star–delta starter, the interlock is used to prevent and lock the handle in place if the operator has left it too late to change over from star to delta, as it would cause current surge as speed would have dropped off. This surge could also cause voltage dip

6. With a star – delta starter it is necessary to change over when the motor has reached its normal running speed.

7. The voltage tappings with an auto transformer are for different starting conditions, ie they normally have three such tappings, such as 40%, 60% and 75% depending on the starting torques.

8. With an auto transformer starter, to overcome the transition switching period, a method called the closed transition 'Kornorfor' starting method is used. (A set of relays and timers are used).

9. The kind of starters now becoming more common at sea are electronic starters, often referred to as 'soft starts'. Thyristors or a combination of thyristors and diodes are used to control the current flow during motor starting.

9.4 Answers – Shore Supply

1. Shore supply is required when the ship's generators and their prime movers are shut down for major overhaul during a drydock period.

2. You cannot normally parallel shore supply with the ship's generators.

3. To correct phase sequence, any two phases are swapped over.

4. If a 60hz ship is supplied with a 50hz supply from shore, the voltage reduction is 17% lower.

9.5 Answers – Electrical (General)

1. The four kinds of switchboard are main switchboards, emergency switchboards, section boards and distribution boards.

2. The instruments found on a switchboard are synchroscope, ammeter, voltmeter, wattmeter, frequency meter and hour meters.

3. To dress contacts on a circuit breaker, a smooth file should be used. Carborundum or emery cloth should not be used as hard particles can be embedded in the soft copper contacts and cause future troubles.

4. The types of closing mechanisms used in circuit breakers are:

    - Independent manual spring – the spring charge is directly applied by a manual depression of the closing handle
    - motor wound stored charge spring – closing springs are charged by a motor/gearbox unit. Spring recharging is automatic following closure of the breaker. Breaker operation is by button, this may be direct manual release of the charged spring or via electrical release solenoid latch
    - hand wound charged spring – similar to the above, but with a manually charged closing spring
    - solenoid (hold on contacts – the breaker is closed by a DC solenoid energised from the generator or bus bars via a transformer/rectifier unit, contactor, push button and sometimes a time relay.

5. A circuit breaker may be caused to trip by:

    - Manual operation (push button)
    - undervoltage trip coil (trips when de-energised)
    - overcurrent/short circuit trip device (trips when energised)
    - solenoid trip coil - when energised by a remote switch or relay such as an electronic overcurrent relay.

6. The typical insulation resistance of an MCCB is 4 – 5 megohms.

7. The current range used with MCBs is 5 – 100 amps.

8. The instruments used for parallel operation of a generators are a synchroscope, voltmeters, frequency meters and wattmeters.

9. The voltmeter measures the potential difference between two points in the circuit, for paralleling two are required, one to measure the voltage across the busbars and the other to measure the voltage of the incoming generator.

10. The currents normally found in the secondary winding of a current transformer are 1 – 5 amps.

11. The most common form of ammeter or voltmeter is the moving coil, iron type or repulsion type.

12. The kind of instrument used for a voltmeter is usually the moving iron type.

13. The phase to which synchronising lamps are attached are as follows; the key lamp is connected in one phase but the other two are cross connected.

Sketch of synchronising lamps.

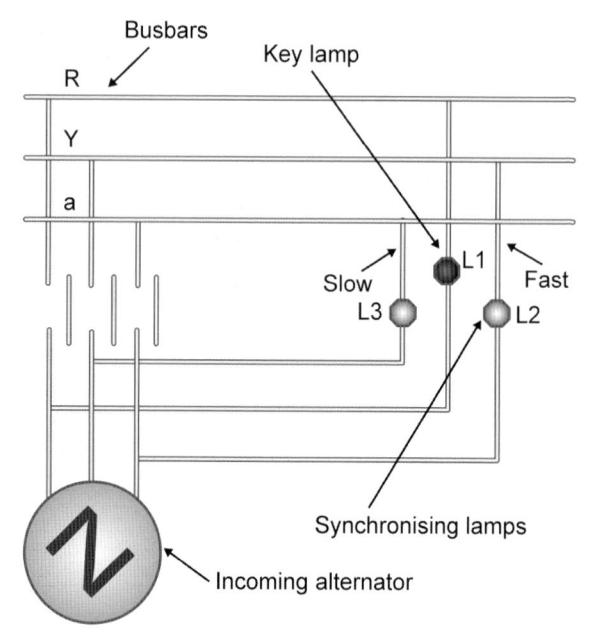

Figure labels: Busbars; Key lamp; R; Y; a; Slow; L1; Fast; L3; L2; Synchronising lamps; Incoming alternator

14. The problem with paralleling AC alternators is that the two voltages must be the same and remain the same after paralleling.

15. The operations necessary for paralleling are:

- The speed of the incoming machine must be adjusted until its frequency is approximately equal to that of the generators that are already connected to the busbars
- the voltage must be adjusted to correspond with that of the busbars
- the incoming machine circuit breaker must be closed as near as possible to the moment when the instantaneous voltages are in phase (ie phase angle zero) and in equal magnitude.

16. The synchroscope must be moving in the fast direction (clockwise) so that the incoming machine will immediately assume load. If the machine was switched when running slow, it would take a motoring load which might possibly operate the reverse power relay.

17. When using a synchroscope, the correct moment to close the breaker is when the needle is moving in the fast direction around 1 rev/5 seconds and when at the eleven o' clock position, the breaker is closed.

18. The typical droop on an alternator governor is usually 2 – 4%.

19. To prevent an error in synchronising, a check sychroniser is used.

20. A good check synchroniser monitors phase angle, voltage and frequency and only allows closing of the circuit breaker when preset parameters are met.

21. It is undesirable to make repeated successive starts of a motor as it will cause the motor to overheat and damage it.

22. The disadvantage of a DOL starter is that it has a high initial starting current (5 – 8 times full load current).

23. The starting torque of a motor in star–delta starters is around 80% of full load torque.

24. The kind of starter used for starting large motors is an auto transformer starter or a star–delta starter.

25. The minimum period of cooling after two consecutive starts with an auto transformer starter is 60 minutes.

26. Intermittent duty with regard to starters means the types of starters used for frequent starts (40 starts per hour).

Engineering Pocket Book – Answers

27. Electronic starters, often referred to as 'soft starters', use solid state technology (thyristors or a combination of thyristors and diodes) to control the current flow during starting.

28. The adjustments that can be made to an electronic starter are:

- Voltage ramp – this sets the time for the motor to reach full voltage output
- current limit – this adjustment is used to prevent the starting current exceeding a preset limit
- initial firing angle – it is often important that the drive should start as soon as the voltage is applied, ie if the drive is standby to a duty.

29. When motors run at a higher frequency, ie if a 50hz motor was connected to a 60hz supply, they would run at 20% overspeed which would also effect the starting torque.

30. If a 50hz ship is supplied with 60hz, the increase in centrifugal loads would be 73%.

31. If a motor stalls, this is normally due to lower than normal frequencies which cause them to run slower and overheat.

32. Heating of a motor varies in proportion to the current, in which heating is proportional to the square of the current, ie, motor heating with seven times the full load current is forty nine times full load heating.

10 ELECTRICAL PROTECTION

10.1 Answers – Electrical Protection (General)

1. Special precautions must be taken in voltages over 55V.

2. High voltage systems are more dangerous than low/medium ones because some high voltage systems are able to retain a lethal charge even when switched off. In addition, dangerous potentials can exist some distance between/from live high voltage conductors.

10.2 Answers – Fault Protection

1. The most serious hazard at sea in electrical systems or caused by electrical systems is fire.

2. The maximum temperature allowable with rated full load current is 80°C (ie 40°C rise above ambient of 40°C).

3. Insulation at high temperature starts to break down, loses its properties of insulation and becomes burnt out.

10.3 Answers – Circuit Breaker

1. The principal factors to be considered when selecting a circuit breaker are system voltage, rated load current and the fault value at the point of installation.

2. Extra factors to be considered with high voltage systems are the methods of earthing.

3. The factors that determine the current rating of a breaker are:

    - The maximum continuous permissible operating temperature of the circuit breaker copper work and contacts

Engineering Pocket Book – Answers

- the ambient temperature
- the temperature rise of copperwork due to the load current.

4. The percentage that the rating of a circuit breaker would change from its free air value due to mounting in a switchboard would be around 80 or 90%.

5. The primary function of a circuit breaker is to isolate any fault that may occur in an electrical system.

6. The four fault ratings of a circuit breaker are:

    - The symmetrical breaking current – the rms value of the AC component of the breaking current
    - the asymmetrical breaking current – the rms value of the total breaking current which includes both AC and the DC components
    - the making current – the peak value of maximum current loop, including the DC component in any pole during the first cycle of current when the circuit is closed
    - the short time rating – the rms value of the current that a circuit breaker is capable of carrying for the started time.

7. If the circuit breaker was rated at less than the expected fault level, the breaker would be liable to explode and cause fire.

8. The size of the short circuit current is determined by the total impedance of the generators, cables and transformers in the circuit between the generator and the fault.

10.4 Answers – Overcurrent Protection

1. System protection discrimination is provided so the generator breaker only trips if a feeder breaker fails to do so or a busbar fault occurs.

Engineering Pocket Book – Answers

2. The problem that can occur if the full load current is greater than the reset current is that the overcurrent device may remain in a partially operated state. From this position, it may creep even with normal load and trip the breaker.

3. The fault current can be higher when the alternator is cold because the field resistance will be low then.

4. In practice, the time delay from 10% full load currents is 20 seconds.

5. The usual short circuit condition time delay for alternator overcurrent protection is 0.1 and 3 seconds, the actual setting depending on discrimination requirments.

6. The common characteristic of all overcurrent relays is the bigger the current, the faster it will operate.

7. Higher viscosity oil is used in marine dashpots because of the higher ambient temperatures.

8. The electronic overcurrent relay operates on the principal that it converts the current into a proportional voltage, this is compared with a set voltage level within a monitoring unit. The time delay is taken by the time to charge the capacitor. This sort of relay usually has separate adjustments for current trip level and for time trip.

9. The type of overcurrent protection found in MCCBs and MCBs are thermal relays.

10. Overcurrent protection is tested by injecting test currents into the breaker to check their current trip levels and time lags. The test is essentially a transformer and controller rather like a welding set.

Engineering Pocket Book – Answers

10.5 Answers – Fuses

1. The fuse size is specified by the motor manufacturer.

2. The kind of fuse not recommended for use at sea is the rewireable type.

3. The advantage of a fuse is its very high speed of operation (a few milliseconds).

4. Before replacing a fuse, the supply must be switched off. The cause of the fault must be located and repaired. The replacement fuse link must be of the correct current rating, grade and type. All three fuses in a three phase supply should be replaced even if only one is found to have blown after a fault.

5. The symbol on an HRC fuse link represents 'high rupture capacity'.

10.6 Answers – Protection Discrimination

1. Protection discrimination is achieved by co-ordinating the current ratings and time delays of fuses and overcurrent relays used between the generator and load. The devices nearest the load have the lowest current rating and shortest operating time and those nearest the generator have the highest current rating and the longest operating time. This will allow the protection system to disconnect only faulty circuits and maintain the electrical supply to healthy ones.

2. The course of action to be taken when a generator is overloaded is that the preference trips should be operated. These will normally operate automatically by means of a timing relay, disconnecting non-essential loads in a definite order and at definite time intervals.

Example: 1st trip Air conditioning, galley, laundry, ventilation

2nd trip Refrigeration plant

3rd trip Deck equipment.

3. Apart from overload, preference tripping may be initiated by low generator frequency.

4. Preference tripping is tested by current injection.

5. In alternators above 5000 kVA temperature detectors are required in the machine windings to comply with certain statutory requirements.

10.7 Answers – Loss of Excitation

1. When there is loss of excitation in a parallel generator, the system voltage may fall slightly, if at all. This is because the excitation on the other machine or machines will increase and offset any tendency for the voltage to fall. The faulty machine will then continue to supply kilowatts by operating as an induction generator. Unfortunately, this current is seen as a large circulating current flowing between the faulty and healthy machine and could eventually cause damage to the faulty machine, but of more importance, cause tripping of the healthy machine. To prevent this, relays are fitted (excitation loss relay) and are normally set below the normal overcurrent relays, 75 – 100 % of full load current being typical. Similarly, the time delay relay of the excitation loss relay should be shorter than the overcurrent operating delay.

2. In the event of the above occurring, the alternator that would trip would be the one with the loss of excitation and the relay that would operate would be the loss of excitation relay.

10.8 Answers – Undervoltage Protection

1. The undervoltage trip will operate at around 70% value or less.

2. Motor starters are fitted with undervoltage trips to prevent the generators from tripping when power is restored owing to the total starting current of all connected motors. Some motors have special automatic restarting facilities.

3. The kind of relays used for undervoltage protection are time delayed relays to permit the prior operation of a feeder breaker.

10.9 Answers – Reverse Power Protection

1. The reverse power relay monitors the direction of power flowing between the generator and the switchboard.

2. A steam turbine overspeed trip is connected to the circuit breaker trip because very little reverse power (as little as 3%) is required to cause overspeeding.

3. In diesel generator sets, the extra load that occurs when one set motors is load on the remaining set or sets.

4. The reverse power delay setting for steam turbine sets is between 1.5 – 3% of full load settings.

5. There is a delay on the relay to prevent inadvertent operation of the reverse power trip due to power surges, particularly during synchronising.

6. Instead of reverse power protection, electrical interlocks or contacts can be used which will respond to various conditions such as closing of the fuel or steam admission valve.

10.10 Answers – Motor Overload Protection

1. The most common type of overload protection is the bimetallic thermal overload relay.

2. The advantage of a thermal relay over a magnetic relay is that it can be approximately designed to follow the motor heating curve.

3. The disadvantage of thermal relays is that when circumstances require, a motor cannot be restarted immediately until the bimetallic strips have cooled and reset.

4. Magnetic relay overloads are of the solenoid type, consisting of an iron plunger surrounded by a coil, one for each phase protected. The coil carries line current, or a proportion of line current if transformers are used. At a predetermined value, the plunger is attracted to the centre of the coil and a pushrod activates a trip bar which opens an auxiliary contact in the operating circuit and activates a mechanical latch.

5. The delay in a mechanical relay can be effected by an oil dashpot, the time lag on all oil dashpots being dependent on the viscosity of the oil used, which is further affected by temperature.

6. Built-in motor overload protection can be achieved by using thermisters.

7. Primarily using thermistors protects against the causes of motor overheating which are not reflected in the motor current, ie blocked ventilation.

8. The advantage of a thermostat over a thermistor is that they are not as bulky or as expensive and they can operate on control gear without additional equipment.

Engineering Pocket Book – Answers

9. The features that high specification electronic relays can provide are:

    - Thermal overload protection with adjustable time/current characteristics
    - overload alarm
    - high set overcurrent protection
    - zero phase sequence or earth fault protection
    - negative phase sequence or phase unbalanced protection
    - undercurrent protection.

10.11 Answers – Earth Faults

1. One earth fault in an earthed distribution system would be equivalent to a short circuit fault across the load via the ship's hull, the resulting large earth fault blowing the fuses in the line conductors.

2. The distribution system that is more efficient at maintaining a supply is the insulated distribution system (used in most marine electrical supplies). If one earth fault occurred in this system, it would not cause any protective gear to operate and the system would continue to function normally. Important equipment still operates. The single earth fault does not provide a complete circuit, so no earth fault current can exist. It would take two earth faults on two different lines to cause an earth fault current.

3. When considering an earthing resistor, the ohmic value is usually chosen so as to limit the maximum earth fault current to not more than the generator full load current.

4. The maximum voltage of electrical systems in tanker cargo areas is 3 phase 440V.

Engineering Pocket Book – Answers

5. The sign of an earth fault using earth lamps is as follows:

    - If the system is healthy (no earth faults), the earth lamps all glow at equal half brilliance
    - if an earth fault occurs on one line, the lamp connected to that line is dim or extinguished and the other two glow.

    Sketch of earth lamps.

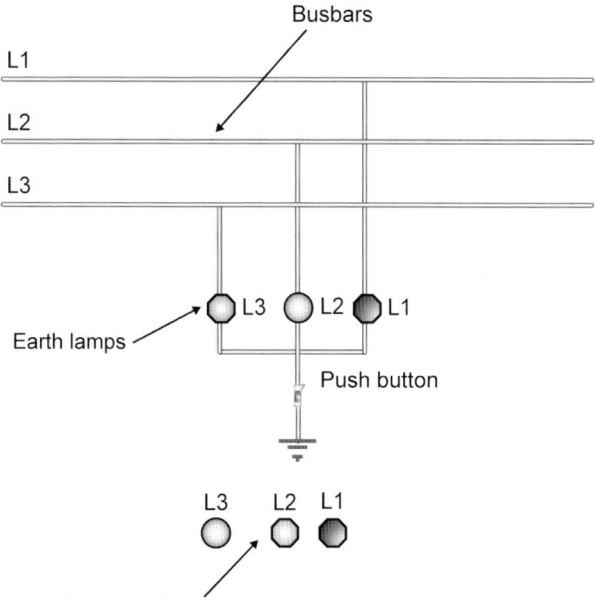

With the push button closed the earth lamp indication shows as earth on L1

6. The indicators an earth fault instrument give are a visual and audible indication in the event of an earth fault. If healthy, insulation resistance is high and greater than 1 M ohm.

    If insulation is faulty, insulation resistance is low and less than 5 M ohms.

11 ELECTRICAL SAFETY

11.1 Answers – General Electrical Maintenance

1. Alternator windings should be checked for signs of oil or water contamination and any damage.

2. Alternator terminal boxes should be checked for tightness of terminal connections and also for damaged or frayed insulation.

3. Alternator air-ducting should be checked for blockages and that it is free from dirt and dust.

4. When cleaning windings with low-pressure air, care should be taken that dirt is not driven deeper into the windings.

5. Alternator slip rings should be checked for even wear and that carbon brushes have free movement in their boxes. Correct brush pressure can be checked a pull type spring balance (normal pull around 1 – 1.5 kg is usual or according to manufacturer's instructions). Minimum brush lengths should be no less than 2 cm or spring pressure will be too little, which could cause sparking.

6. If winding resistance is less than 0.5 M ohms, they should be given a thorough cleaning and dried out. If the value has recovered to a reasonable value which has become steady during the drying out period, its windings should then be coated with high quality air drying insulating varnish. Should the IR values remain low, this could mean the machine needs to be rewound.

7. If an alternator is to be laid up for a considerable period, ensure that the windings are suitably heated to prevent internal condensation forming on its insulation.

11.2 Answers – Switchboard Maintenance

1. Precautions to be taken when working on switch gear are that it is isolated and to ensure that it cannot be made live. Where interlock circuits, pilot lights and control circuits are involved, they may be supplied from a different source, so this must be checked as well.

2. Before starting work, the actual equipment should be checked with a live line detector, not only between phases but also between phases and earth.

3. A switch gear insulator should be checked for signs of tracking and blistering in the vicinity of exposed live metal. If the material is of the bonded laminated type, the laminate should be checked for signs of spills along it.

4. Alignment of contacts is checked by removing the arc chutes on ACBs to expose the contact assembly. On some high voltage designs, the chutes tilt forward to provide access.

5. To dress copper contacts, a fine file or fine glass paper can be used.

6. The maintenance carried out on silver plated contacts is very little but if cleaning is necessary, metal polish may be used.

7. Excessive application of petroleum jelly can cause burning and pitting of contacts.

8. Dash pots are checked for correct oil levels. Particular care should be taken to use the correct grade of oil when topping up or replenishing, as proper operation of switch gear is entirely dependent on this.

9. Trips and relays are checked by using a current injection tester.

11.3 Answers – Motor maintenance

1&2. Dirt on motor insulation causes two things, firstly it may in time cause carbonisation of the deposit and of the surface of the insulation, and eventually a burn out. Secondly, the vent ducts may become clogged, causing overheating and failure.

3. In maintenance, cleanliness is next to godliness.

4. With totally enclosed motors, a layer of dust can cause overheating as heat generated is removed through surfaces and the dust acts as an insulator.

5. Air pressure must be less than 1.75 bar when blowing out motors so that dirt is not driven further into the windings.

6. A source of oil and grease contamination in motors can be from motor bearings.

7. Oil and grease contamination should be remove from a motor by brushing and cleaning with one of the many recommended brands of cleaning fluids. Broken or missing bearing covers must be replaced to prevent grease escaping.

8. When a motor is dismantled for cleaning, it must be inspected thoroughly. This way, faults can be detected before they lead to breakdowns.

9. Damaged insulation of the stator windings could be caused by careless replacement of the rotor into the stator.

10. Discolouring of the stator winding's insulation could be caused by overheating. The fault must be found and remedied.

11. Rubbing of the stator core could be caused by a worn bearing.

12. After changing bearings on a motor, check for signs of the motor overheating and the current each phase is drawing. If

Engineering Pocket Book – Answers

there is a sign of overheating, the motor should be dismantled and core checked for signs of discolouration, which will indicate overheating.

13. Moisture is detected in motor windings by taking insulation readings.

14. Bearings should be renewed as a planned maintenance policy, or as per manufacturer's instructions.

15&16. The procedure for a bearing that cannot be replaced is that it should be cleaned by being immersed in a solvent such as clean white spirit or paraffin, then cleaned in a jet of clean, dry compressed air. Once dried, the bearings must be lightly oiled. Any traces of metal particles from the bearing indicates wear or damage and the bearing must be discarded. Also carefully examine raceways and rolling elements for signs of wear and damage.

17. If a bearing sticks, it must be rewashed and inspected. If it sticks after this, it should be discarded.

18. Visual signs that a bearing should be scrapped are damaged raceways or roller elements, any signs of metal particles from the bearing or sticking bearings.

19. The amount of grease that should be put in a motor bearing should be one third to half full, as over greasing causes churning and friction, which results in heating.

20. If a bearing housing has no vent holes, before regreasing, the housing will have to be cleaned out before charging with fresh grease.

21. The application that requires special greases for motor bearings is when the motors are used for fan motors in refrigerated compartments where they must operate at freezing temperatures.

22. A motor is inspected for signs of damage and overheating in the cage and laminated core. Also make sure all core ventilating ducts are clear.

23. Salt contamination of a motor is removed by cleaning, drying and then revarnishing.

24. Stator windings are usually dried out with low power electric heaters or lamps, with plenty of ventilation to allow dampness to escape.

25. Tests carried out before revarnishing are IR tests and if they remain high for a couple of hours, the motor is then revarnished.

11.4 Answers – Motor Control Gear Maintenance

1. All moving contacts used in control gear have what is known as a wipe or follow through (a cleaning action).

2. If the wipe action is lost from moving contacts, a build-up of oxide could occur on the contact surfaces and cause overheating.

3. The condition contacts have to be in before filing are badly burnt or pitted, then a fine file should only be used sparingly.

4. Mechanical wear can be reduced by applying a thin layer of oil on the contacts.

5. Overheating in copper contacts can be caused by an oxide film forming on them (due to being closed for long periods).

6. Rust is removed from a magnet by using fine emery cloth.

7. Starter enclosures should be checked for accumulation of dirt or dust, corroded parts, fixing bolts and earth bonding connections.

8. In high vibration areas, fixing bolts, earth bonding connections and all terminal connections should be checked.

9. Fixed and moving contacts should be replaced in pairs.

11.5 Answers – Maintenance of Lighting

1. When replacing lamps or tubes, it must be ensured that the supply is off.

2. Arc flash can cause personnel blindness, burns and can cause fires.

3. If a higher than rated light is fitted in a lamp fitting, it can cause overheating and fire.

4. When changing fluorescent tubes, it must be remembered that the capacitor in the circuit may retain a charge and should be discharged before handling its terminals.

5. With portable cargo lighting, check that cables and supply plugs are in good condition.

6. Corrosion in flameproof light enclosures can reduce their strength.

7. If the lamp glass is cracked in a flameproof light fitting, a complete new lamp glass assembly should be fitted immediately.

8. If bolts are overtightened in a flameproof fitting, it could cause distortion of the flame path, damage seals and cause excessive stress on lamp glasses.

11.6 Answers – Electric Shock

1. The maximum current for 'let go' is 13 – 15 mA (greater for DC).

2. The maximum voltage regarded as reasonably safe for portable power tools is below 60V.

3. The maximum shock voltage to earth for a centre tapped 110V tool transformer is 55V.

4. The first thing that must be done when finding a victim of electric shock is to detach them from the live conductor.

5. Two methods of doing this are to switch off the supply or remove the victim using a dry, non-conducting material.

6. Details of first aid to be carried out in the event of electric shock can be found on switchboards, high voltage equipment, etc.

11.7 Answers – Circuit Safety Devices

1. Circuit safety devices fitted to switchboards are circuit breakers, fuses, reverse relays and current sensitive relays to provide protection against faults in the distribution system.

2. Under fault conditions, the last protection device to operate should be the generator circuit breaker.

3. All fuses should be of the non-renewable cartridge type and capable of interrupting fault current.

4. Two methods of tripping non-vital loads to ensure essential supplies are automatic tripping of all non-vital supplies or have a timed sequence of tripping of non-vital loads until the overload is reduced to an acceptable value.

11.8 Answers – Earth Safety Devices

1. An earth fault is stopped at a transformer by the air gap between the windings.

2. An earth return path must have low impedance so all of the earth fault circuit has low impedance.

3. The value that earth fault leakage breakers are set to trip at is at currents as small as 40 – 50 mA.

11.9 Answers – Welding

1. The normal range of welding current is from 50 – 600 Amps.

2. The two main non-electrical safety precautions used to protect welders are protective clothing and visor.

3. To protect operators from indirect electrical shock, the work piece should be well earthed.

4. The type 'B' electrode holder is better than the type 'A' because no live parts are accessible to the standard test finger and when the electrode is fitted, its non-coated end does not protrude from the electrode holder.

11.10 Answers – Static Electricity and Interference

1. Equipment that suffers from interference is classified as susceptible.

2. An example of non-electrical equipment giving rise to static is a shaft on the rubbing face of a bearing.

3. An example of a very common source of interference which is a steady state condition could be from a fluorescent tube, generator or motor.

4. Intermittent interference could be caused by erratic connections.

5. The only way of guaranteeing the prevention of interference is by keeping equipment that causes interference as far away as possible but this is not practical or possible on a ship.

6. Radio transmitters and receivers are constructed so that their casings form complete electromagnetic screens to prevent interference.

7. Suppressors are fitted in power supply lines as close as possible (not more than 150 mm away) to the equipment being protected.

8. Cables that should not be run together are sensitive and non-sensitive cables.

11.11 Answers – Emergency Power Supplies

1. An emergency electrical power supply must be provided so that in the event of an emergency involving total power failure, a supply will be available for emergency lighting, alarms, communications, watertight doors and other necessary services to maintain safety and to permit safe evacuation of the vessel to lifeboats.

2. An emergency generator must be provided with its own IC engine as a prime mover, its own fuel supply tank, starting equipment and switchboard, and must be above the water level.

3. An emergency power source must come on line following a total electrical failure.

4. Both an emergency generator and batteries are recommended so that at the moment of total power failure, there is not a black out, as a battery supply will instantaneously supply power to lighting.

5. On passenger vessels, battery powered lighting must be able to run for a minimum of 3 hours.

6. Testing of emergency power supplies are done weekly on a normal basis where practical.

Engineering Pocket Book – Answers

7. The special interlocks that are fitted to the main and emergency circuit breakers are interlocks that prevent parallel running of the emergency and ship's generators as it is not normally possible to synchronise them.

8. The two most common methods of automatically starting an emergency generator are by compressed air and battery.

9. The auto-start can be tested by simulating a power failure by pulling a fuse in the auto-start panel which supplies the undervoltage or under frequency relay.

10. The only way to test the performance of an emergency generator is to give it a proper load test by means of disconnection of normal mains power.

11. Equipment that is normally fed by batteries is emergency lighting, emergency generator starting, radio equipment, navigation lighting, alarm panels and communications.

12. Two types of battery cell are the lead acid and the alkaline cell.

13. The range of voltages of a lead acid battery cell are 2 volts.

14. Battery capacity is rated in ampere hours (Ah).

15. During charging of a battery cell, the gas given off is hydrogen.

16. The two types of cell must be kept separate because rapid electrolyte corrosion to metal work can take place and certain damage to batteries can occur.

17. The electrolyte's SG in lead acid batteries indicates the charged state of the battery, ie SG of 1270 – 1285 when fully charged and 1100 when discharged.

18. The arrangement used in battery charging equipment is normally a transformer/rectifier unit to supply the required DC voltage to the cells.

19. The maximum allowable electrolyte temperature during charging is 45°C.

20. The only indication of a fully charged alkaline cell is when the voltage remains at a steady maximum of 1.6 – 1.8 volts.

21. If the cell plates are exposed to air, the life of the battery will rapidly reduce.

12 INSTRUMENTATION AND CONTROL

12.1 Answers – Control (General)

1. The two major factors that have influenced the development of control in the marine environment are economy and human factors, ie wages and reduced manning.

2. Twelve main parameters monitored at sea are pressure, temperature, level, flow, viscosity, speed, torque, voltage, current, power, machinery status and equipment status.

3. The UMS class notation means unattended machinery space, where vessels are capable of safe operation for any period of time, usually eight hours maximum, without being manned.

4. CCS class notation stands for centralised control station where control and monitoring equipment can be grouped together. This option of operation qualifies the vessel for the class notation CCS.

5. The five different control media are mechanical, hydraulic, pneumatic, electric and electronic.

6. The process of long distance control transmission is called telemetry.

7. The modern definition of a transducer is a device that converts a signal in one medium to a signal in another, such as a pneumatic to electric.

12.2 Answers – Control Systems

1. A control system is a group of separate components or sub systems that are connected together in order to regulate a variable quantity or several variable quantities simultaneously.

2. Control systems are classified as open loop or closed loop systems.

3. Information to the controller is provided by feedback.

4. The term 'transmitter' is commonly used to describe the measurement element in a control system.

5. The behaviour and performance of a control system depends upon the interaction of all the elements in the system, ie the best controller in the world cannot make an adequate plant work well.

6. If a control system's action is dependent upon the output, it is called a closed loop or feedback control system.

7. Excessive corrective action is called unstable.

8. The difference between input and output signals is called or determined as the error.

9. The three common essential elements of a control system are a measuring device, a controller and an actuator.

10. The three components of a transducer are a sensing element, a conversion device and compensating arrangement.

11. The set point value is the value to be obtained or kept by the controller.

12.3 Answers – Measuring Instruments

1. The three methods of measuring pressure are:

   - Balancing the unknown pressure against a column of liquid with known density
   - by elevating the stress in an elastic medium caused by the unknown pressure
   - by balancing the unknown pressure against a known force.

2. The six factors affecting the deflection of a bourdon tube gauge are dependent on the radius of the tube, the total length of the tube, the wall thickness, cross sectional area of tube and Young's modulus of the material.

3. The five factors affecting the deflection of a diaphragm sensing element are the radius of the element, the number of corrugations, the thickness, the depth of each corrugation and young's modulus of material.

4. To increase the spring rate of a diaphragm capsule, a spring is used.

5. The materials that diaphragm sensing elements are normally made from are brass, phosphor bronze, beryllium copper or stainless steel.

6. The method usually used at sea for level measurement is by instruments using static head measurements.

7. A second liquid is sometimes used in a single column level indicator because it may be convenient to use a different liquid of different density as the height may be too great for use of a sight glass.

8. A diaphragm level indicator must be mounted clear of sludge at the tank base.

9. A dip tube level indicator is used when the bottom access to a tank is not practical or available.

10. To gauge tank level with a dip tube level indicator, the head of tank pressure is measured.

Engineering Pocket Book – Answers

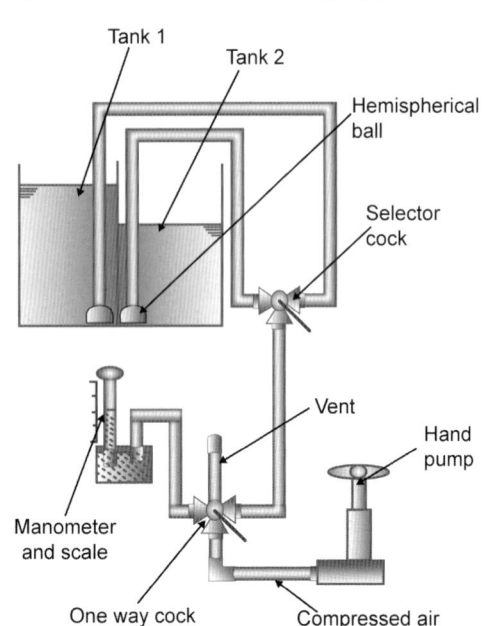

11. Operation of a Pneumercator tank gauge, see sketch below

To use the system above, the gauge cock is set at pressurise and the hand pump operated to charge the balance chamber in the tank. The control valve is then set to gauge and the reading noted. This operation is repeated several times until two readings the same are recorded and surplus air pressure has bled from the chamber via a notch so the back pressure in the chamber equals that of the static head.

12. A mercury trap is fitted to the manometer so mercury is not blown in the event of a blockage.

13. The bell of this gauge must be clear of any sludge level in the tank.

14. The minimum lengths of straight pipe before and after an orifice plate must be ten pipe diameters of length upstream and five pipe diameters downstream.

15. The upstream straight section can be increased if there are any bends, double bends or partly opened globe valves upstream of the orifice.

16. The leading edge of the orifice plate must be sharp and right-angled with no burrs on the edge.

17. The type of pressure tappings that are used for an orifice with small-bore pipe are corner tappings or flange tappings.

18. When measuring fuel oil flow with an orifice plate, seal pots are fitted between the differential pressure instrument and the pressure tappings to prevent coagulating the instrument with fuel oil.

19. Before zeroing a control instrument transmitter, the control system must be put on manual control.

20. The variable area flowmeter works by having a float move up or down in a chamber in proportion to liquid flow and beside the chamber is a calibrated scale with flow on it.

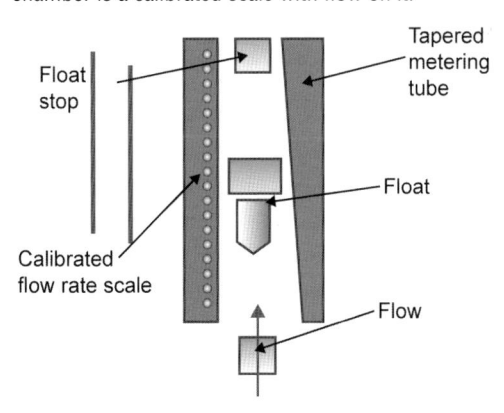

21. The minimum flow that can be monitored through a variable area flowmeter depends on the weight of the float.

23. To calibrate a turbine flowmeter, the guide vanes on the upstream side are adjustable.

24. Possibly the most important variable measured at sea is temperature.

25. Errors are caused in temperature measuring systems because often insulating materials are present between the sensing element and the medium to be measured.

26. A bi-metallic thermometer works on the principal that it comprises of a bi-metallic strip of metal formed into a helix such that when the bi-metallic strip comes into variation of temperature, the two different coefficients of expansion will cause the helix to rotate, operating a mechanical linkage, providing an indication of temperature on a scale.

See sketch

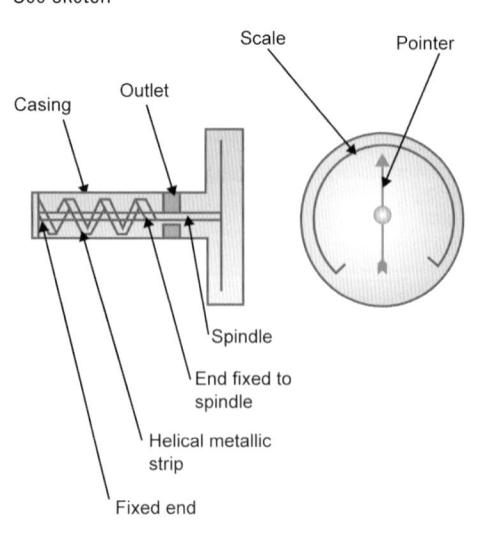

27. The most common form of bourdon tube in mercury-in-steel thermometer is a type of bourdon tube of almost flat cross section, wound in the form of a helix.

28. The most popular type of distance reading thermometer is a force balance temperature transmitter having a gas-filled element.

29. Helium and nitrogen are popular for gas-filled thermometers because of their inert character.

30. The main disadvantage of a gas thermometer is that through time the gas can diffuse slowly through the metal bulb and capillary, particularly at welds.

31. A resistance thermometer temperature element is constructed by means of the resistance bulb being wound from platinum or nickel wire of high purity on a mica or ceramic former and the complete spool is often glazed to protect the wire from contamination.

32. The temperature element can be sealed in a small diameter stainless steel sheath, evacuated and filled with hydrogen which is a good thermal conductor (best of all gases).

33. In a resistance thermometer, the bridge resistances are made up of manganin wire, which has a very low coefficient of resistance.

34. To eliminate the error caused by ambient temperature in a two wire resistance thermometer, 3 and 4 wire connecting lead systems are used, as shown below.

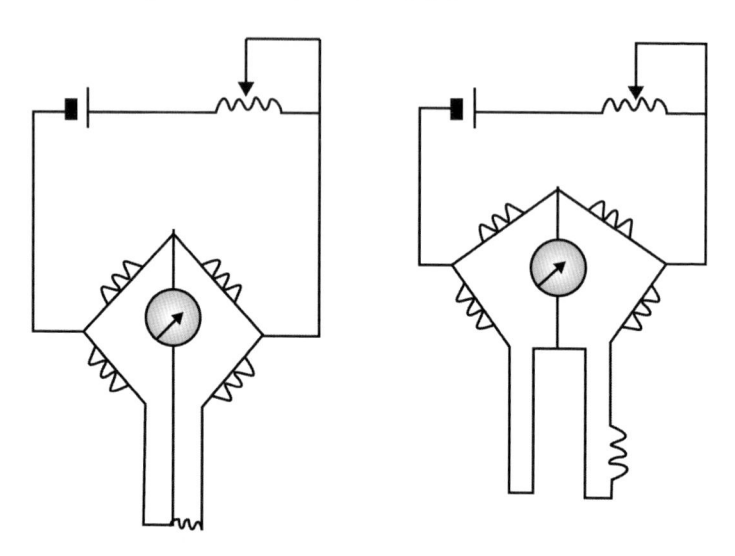

35. The most popular method is the three-wire method.

36. High bridge current in a resistance thermometer can cause a possible heating effect, causing changes in bulb resistance internally.

37. The type of bridge that is more accurate in a resistance thermometer is a balanced bridge type.

38. The cold junction of a thermocouple is connected to the measuring instrument by compensating cables which have the same thermo-electric properties as the thermocouple wire but over a lower temperature range (0-100°C).

39. Usually the minimum range of a moving coil millivoltmeter is 8 mV (approx 0-200°C).

40. The most common method of eliminating the effect of ambient temperature from a moving coil temperature indicator is by fitting a bimetallic spiral in the instrument to the top hair spring and in opposition to it, so that as the temperature varies, the free end of the bimetallic spiral adjusts the pointer position to take this into account.

41. The three methods of connecting thermocouples are:

- Averaging – by this method below, the average temperature can be measured and any number of thermocouples may be used

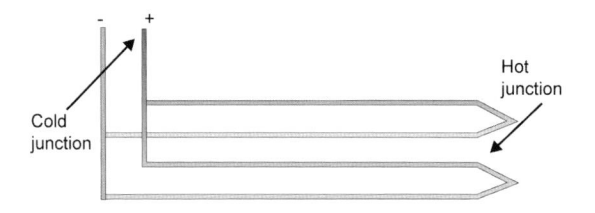

- addition – in this way a number of emfs can be summated when, for example, the output of an individual thermocouple would be difficult to measure. It is important that the cold junctions are all kept at the same temperature

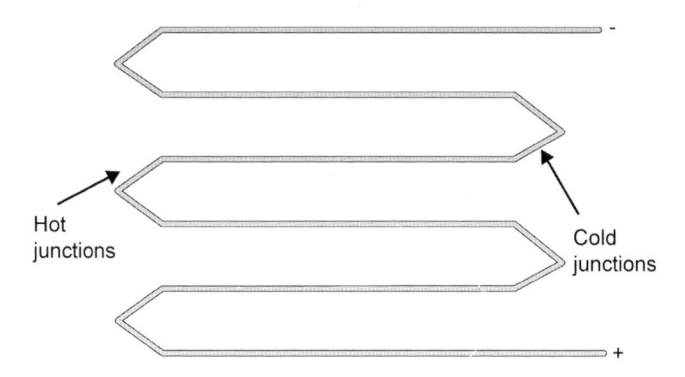

- subtraction (differential) – by this means, temperature rise, in say a heat exchanger, can be measured directly.

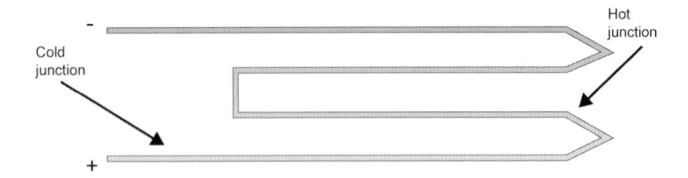

42. The fastest possible response from a thermocouple is obtained by having the tip in good thermal contact with the bottom of the thermowell and this means that there will be an earth return back to the instrument.

12.4 Answers – Automatic Control Theory

1. Feedback is the transmission of a signal which represents the controlled condition for comparison with a signal preset by the operator, and which is used to determine the value of the controlled condition.

2. Open loop control is where there is no feedback of information on the value of the controlled condition.

3. An example of two step control is a level controller for a boiler, ie if the boiler water level were to drop due to an increased demand in steam by 20 mm, then the controller output to the feedwater valve may increase from 0.6 to 0.7 bar. If it then dropped by 40 mm, the controller would increase from 0.6 to 0.8 bar.

4. In two step control, you may get oscillation or hunting occurring.

5. Another name for modulating control is continuous control.

6. The basic form of modulating control is proportional control, where the controller is set up so that any change in output is directly proportional to the deviation between the controlled condition and desired value.

7. The best performance of a control system and plant is obtained by the controller being able to make adjustments to suit the plant condition.

8. To maintain control system performance, the proportional band is adjusted.

9. A proportional band of 100% means that it has to take the full input signal to give the full output signal range.

10. The gain of a controller is where the controller has multiplied the input signal.

11. With proportional control, the only time the controller operates is when there is a change in plant loading.

12. The kind of error that is offset is when, for example, a level has changed due to demand, this will not change until there is another demand causing a change. This is known as offset or steady state error.

13. An offset error will remain until a change in demand causes the controller to operate.

14. The difference between desired value and the set point is called offset.

15. The set point would be adjusted to correct the droop of a controller with every load change.

16. The offset is reduced in proportional control by increasing the gain or sensitivity of the controller.

17. Oscillation of the correcting element is called hunting.

18. The name given to integral action is reset action.

19. The other name for instability in a control system is hunting.

20. Hunting is avoided with a proportional controller by changing to hand control to bring the process to a steady position.

21. The reset effect is measured by the number of repeats per minute of the controller.

22. The adjustment of integral action must be gradual otherwise hunting will occur.

23. The name given to derivative action is rate control or three term control.

24. To eliminate hunting when there are long rest times, a wide proportional band is required.

25. Derivative action time is used to stabilise and improve recovery from disturbance in a plant with such problems as long reset times.

26. The integral action time when the integral action is large is proportional to it.

27. The type of control that uses two or more correcting elements for the same controller output is known as split control.

28. To avoid problems with two open control valves, when using split range control, a dead band is used, one valve closing at 0.4 bar for instance and the other opening at 0.5 bar.

29. The type of control that keeps two variables in fixed proportion is ratio control.

30. Cascade control is used when something large is to be controlled, such as a high volume of water or a large thermal

storage capacity, the two controllers may be used in series and is known as cascade control.

31. Three types of delay in system response are distance velocity lag, measurement lag and transfer lag.

32. Distance velocity lag is reduced by placing the sensing element, ie a thermometer, close to a heater or sensor which is sending back the signal.

33. Transfer lag is sometimes known as multi-capacity lag.

12.5 Answers – Types of Control Action

1. The condition of response that is most important in design considerations is the time dependence of a control system's response to change.

2. Most failures in a control system occur outside the controller, ie wire breakage, transducer fault or power failure.

3. A controlled condition could be caused to go from one extreme to another due to loss of feedback signal as a result of transducer failure or wire breakage.

4. A controlled condition could be caused to stay at zero due to failure of the output element, actuator or valve, or its connections or power supply.

5. A controlled condition could be caused to oscillate or hunt by, for example, a hydraulic system having air in the system causing springiness of the drive.

12.6 Answers – Automatic Controllers

1. For the force balance beams illustrated, the comparing elements are the diaphragms 1 and 2 which are fed from the set point and feedback signal.

Engineering Pocket Book – Answers

2. The position of the beam pivot for 100% proportional band is in the middle.

3. Gain is increased by moving the pivot to the right.

4. To eliminate offset, integral action or reset action may be applied.

5. The integral action time for this beam type controller depends on the resistance of the valve setting.

6. To give rate action to the beam type controller illustrated, a second controller is used.

7. To overcome time lag, a third beam is added.

8. To increase reset action, the restricter valve is opened wider.

9. The type of controller most common at sea is the automatic three term controller.

10. For the nozzle/flapper controller illustrated, the input signal alters the output pressure in proportion to the airflow supplied.

11. For minimum feedback, the pivot in the flapper nozzle should be as near to its origin as possible.

12. The bellows that give negative feedback are B1 in the diagram illustrated.

13. The bellows that give positive feedback are B2 in the diagram illustrated.

14. For the stacked type controller, the usual ratio of the areas of the diaphragms are that the smaller ones are half the size of the larger.

15. In the stacked controller illustrated, the response is adjusted to suit plant requirements by adjusting the adjustable restriction valve.

Engineering Pocket Book – Answers

16. The usual current range for an electronic controller is 0 – 10 mA.

17. When an electronic controller amplifier drifts off value, it produces a signal and the control circuit will start to rectify an error that does not exist.

18. AC amplifiers are preferred for electronic controllers because there is less of a problem with drift.

19. An AC amplified signal is used for control purposes of energising transistors and making them operate ON-OFF switches, chopping the DC signal to prevent zero drift.

20. Reset action time is varied on the electronic controller illustrated by adjusting variable resistor R.

21. The difference between the pulse controller and the above types is that this one only has outputs of raise, lower and zero and is not variable.

22. The pulse controller allows, when changing from auto to hand, an instantaneous bumpless transfer.

23. Before making adjustments to a controller, it must first be switched to manual control.

24. On a proportional controller, the proportional band is set by setting to the widest value possible. Switch to auto control according the manufacturer's instructions, and then step up or down the set point and watch for plant response. Return to the original set point and half the proportional band, this is done until 1.5 cycles of plant oscillation is achieved before settling.

25. Optimum proportional band is sometimes called quarter amplitude damping.

26. A two term controller is checked for offset by switching the controller to auto and stepping up or down the set point a small

Engineering Pocket Book – Answers

amount, sufficient to get plant response and repeat procedure for proportional control response.

27. Before settling, the number of oscillations a plant should make is 1.5.

28. After setting a two term controller, the next procedure after the periodic time is to set the reset action time.

29. Before settling, a plant should make no more than 1 oscillation with a two term controller.

30. When a three term controller is set properly, the plant response to change should be set to a minimum depending on operating conditions.

31. A pneumatic remote set point adjuster operates by having an output signal pressure acting on a diaphragm, producing a force equal to the spring force set up by the hand control knob. Changing the spring pressure enables the operator to change the output signal.

12.7 Answers – Control System Components

1. An error detector is a component used to compare the actual outputs of a control system with the desired outputs and generate corresponding error signals.

2. The two classes of error detectors are electrical and mechanical.

3. The mechanical differentials used in rotating systems are used for detecting positional errors.

4. Dead band can be caused in a mechanical differential error detector by backlash in gears.

Engineering Pocket Book – Answers

5. Examples of control systems with positional gyroscopes are ship's steering and stabilisation systems.

6. A positional gyroscope requires periodic resetting because of bearing friction and inbalance can cause changes in initial axial alignment.

7. Examples of rated gyroscope measurements are rate of roll, pitch or yaw of an aircraft or ship.

8. Potentiometers are used for detection of position error in angular or translational position control systems.

9. For error detection in temperature control systems, thermocouples or thermo-resistive elements are used.

10. The four classifications of controllers are mechanical, electrical, hydraulic and pneumatic.

11. The main disadvantage of mechanical error detectors and controllers is that the input and the output of the system must be physically close together.

12. The three types of electric amplifiers used in control systems are electronic, magnetic or rotary amplifiers.

13. Motor speed is regulated in a pump controlled hydraulic system by use of a variable stroke pump which regulates the flow of oil to the motor (servo motor).

14. An example of a controller in a valve-controller hydraulic system is a servo or spool valve.

15. Low pressure pneumatic control systems are used in process control applications.

16. The essential difference between pneumatic and hydraulic control systems is that hydraulic control valves control oil flow and pneumatic control valves control air pressure.

Engineering Pocket Book – Answers

17. The type of pneumatic controller used for displacement measurement is a nozzle/flapper type controller.

18. The type of controller found in high pressure pneumatic control systems is the same type as the hydraulic type spool valve.

19. The control system that uses DC electrical motor output elements is speed control or positional systems where the power requirements are relatively low.

20. The main advantage of field control of DC electrical motor output elements is that small input currents can be used which enable these motors to be energised directly from electronic controllers.

21. The amount of extra pressure that may be required to operate a hydraulic piston actuator may be as much as 30% of the supply pressure in order to overcome resistance to motion.

22. Low pressure pneumatic activators are know as pneumatic motors.

23. Low pressure pneumatic activators are used in the process industry for operating flow control valves.

12.8 Answers – Supply Systems

1. Instrument air is an air supply for instrumentation onboard ship and requires a high degree of solid particle filtration, oil and water removal.

2. A particle as small as 5 microns can block orifices and nozzles.

3. Moisture in a pneumatic system can cause blockage of orifices and nozzles if in droplet form and cause corrosion.

4. Absorption dryers or a combination of both usually dry by means of refrigeration or instrument air.

5. Air is dried in an atmosphere drier by using a solid desiccant such as silicon gel, activated alumna and synthetic zealite to absorb condensation forming on the surfaces of the mentioned materials.

6. The absorption efficiency of a desiccant may be reduced by a high air inlet temperature.

7. The purpose of the air dryer outlet filter is to remove desiccant dust which would collect in pipelines and instrumentation.

8. Dryers are changed over in the system illustrated by a programmed electrical timer actuating a five way solenoid valve which controls air flow to an opposed piston change of mechanism. This is linked to the upper and lower four way valves, changing these over every four hours so that the left hand dryer is reactivated while the right hand dries and vice versa.

9. In the process of regeneration, the drying takes two and a half hours heating and one and a half hours cooling.

10. The auto-unloader's operating air usually comes from the compressor.

11. The purpose of the auto-unloader is to unload the compressor when starting and stopping or to act as a relief valve when continuously running, when maximum pressure has been reached.

12. In a filter regulator, incoming air passes through a mesh filter and then passes up through a regulating valve, the outlet pressure acts on a diaphragm and is counter balanced by an adjustable spring.

13. The filtering of power lines must take place where they enter the equipment cabinet, which must be totally shielded.

Engineering Pocket Book – Answers

12.9 Answers – Practical Control Systems

1. Appreciable economy can be effected on turbine driven ships if the LO temp is controlled at a high temperature in the geared propulsion machinery.

2. For the temperature transmitter in the control system illustrated the transmitter's output signal range is 0.2 – 1 bar.

3. To restore the oil temperature to its desired value, a signal is sent to a control valve in the cooling line, adjusting the cooling water flow and restoring the oil temperature to a desired value.

4. The difference between a temperature control system for a LO and one for boiler FO is that the controller regulates the steam to the fuel oil heat exchanger.

5. For the auxiliary exhaust steam control system illustrated, the steam pressure is converted to a proportional air pressure by the pressure transmitter.

6. With this system, high pressure steam is prevented from being dumped to the condenser by an overlap or dead band of the 3 millibar controller out pressure which ensures both valves never open at the same time, ie one valve is stroked at 0.2 – 0.6 bar and the other at 0.63 - 1 bar.

7. For the simple combustion system illustrated; the pressure transmitter measures the steam drum pressure.

8. The controller output is taken directly to the FO supply valve and the two forced draught controllers.

9. The adjustment made to the forced draught controller to obtain the desired combustion with load fluctuation is to vary the proportional band.

10. Differential pressure across the fuel control valve is maintained by measuring the pressure across the valve using a DP transmitter.

11. For accurate combustion control of high pressure or high capacity boilers it is advantageous to measure steam flow in conjunction with the boiler drum, superheater outlet pressure for accurate control.

12. For the main boiler combustion control system illustrated the type of controller used to receive the steam pressure signal would be a two term steam pressure controller whose desired value would be manually adjusted to the required steam pressure.

13. To combine the steam load, a computing relay is used.

14. Rapid change in steam demand is detected by the steam flow meter.

15. The signal to the burner oil control valve also goes to the FO pressure controller.

16. The source of the measured value to the air flow controller is from the master signal.

17. The desired position of the forced draught fan vanes is 60%.

18. In practice, when these is an increasing load at the forced draught fan, the fan vanes will immediately open up and cater for the new load.

19. Excess air is ensured with a decreasing steam load by having two relays, one for high selector and the other low selector. On decreasing load, the high selector prevents the desired value of the air flow controller being decreased until the fuel load has decreased.

20. The drum water level increases with increased steam demand because pressure in the boiler will tend to drop, causing the release of large quantities of steam bubbles at the heating surfaces. These bubbles force their way through the water in the generator tubes to the steam space in the drum and, in doing so, raise the water level in the boiler drum above that previously indicated in the gauge glass.

21. The type of boilers that single element feed controllers are limited to are boilers where the steam release/water storage ratio is low, ie where the amount of water contained in the boiler relative to steam output is high and also where there are few load fluctuations.

22. Steam flow measurement in itself is insufficient to control the drum water level because of the time lapse between the change in steam flow to the repositioning of the feed water control valve.

23. Steam flow is measured by an orifice plate in the main steam line, between the drum and the primary superheater.

24. The water level controller uses the signal from the master signal for readjustment of the feedwater control valve.

25. For the diesel engine LO system illustrated, the controller output goes via a solenoid valve to the seawater control valve regulating flow through the cooler.

26. The purpose of the solenoid valve in the system is to be used as part of an auto/manual switching system and, when operated, will send the output of the controller and pass the signal from the manual station, which is usually located in the control room.

27. For the jacket water system illustrated, the type of valve actuation used is master and slave.

12.10 Answers – Alarm Indication Systems

1&2. A basic alarm system is a system comprising of a series of pressure or temperature operated electrical switches. When the operating temperature or pressure of a machinery parameter rises or falls outwith a preset limit, the switches closes the circuit to an alarm light and sounds an audible alarm if necessary.

3. After accepting the alarm, it should be immediately investigated and the necessary action taken.

4. The three types of indication on an electrical alarm system are an alarm light, audible and flashing light.

5. The kind of alarm indicator used with pneumatic alarm systems are coloured flags and a whistle.

6. The two scanning speeds of a scanning alarm system are a fast scan speed of 1 point/second or slow scan speed of 1 point/3 seconds.

7. Displayed during the fast scan speed is the associated point number indicated at a rate of 1 point/second.

8. The basic measuring system for temperature is the resistance thermometer.

9. The transducers used for pressure measurement are a barometric capsule or bourdon tube, the free end of which is connected via a mechanical linkage to a wiper which is free to travel along a resistance element.

10. Fitted between a contact alarm and the alarm annunciator is an adjustable delay circuit where the liquid level in tanks is monitored.

11. The three basic functions performed by a scanner at each selected point are:
- Routing the transduced signal into analogue to frequency AF converter

Engineering Pocket Book – Answers

- selecting the program for the point (the type of measurement)
- setting up the point identity display in slow scan mode or enabling the measured value to be updated when the point coincides with that selected on the manual selection matrix, when in fast scan mode.

12. The converter operates by selecting the mode of operation (resistance, thermometer or electrical equipment) and sensitivity, and changes the voltage to frequency where it is sent to a central processor.

13. The converter and central processor are isolated so no interference occurs.

14. The sub-systems of a central processor are the totaliser, the decoder and measured value store and the alarm comparator.

15. The circuitry for each measurement is reset to a reference condition determined by the program before the new measurement.

16. The timing circuit is responsible for the precision timing signals required by the control logic. The control logic governs the sequence of events during the measurement and display cycles.

17. The alarm levels are set on the alarm pin board in decimal form, being converted internally into predetermined code for use in the measuring circuits.

18. When no alarm is detected, the point being scanned further establishes whether a normal condition exists.

19. Groups of annunciators are switched out when machinery is shut down.

20. When the self-check system finds a fault, it will give off its own alarm.

21. The self-check system should be tested once a week.

12.11 Questions – Operation and Maintenance

1. The equipment manufacturer's operating instructions should be followed so that damage does not occur to the instrumentation and control equipment or plant.

2. The kind of transfer that should be obtained when switching from manual to auto is a smooth, 'bumpless' transfer.

3. Control cabinets should be designed with front openings for ease of maintenance.

4. A differential flow type measuring instrument must have a straight length of piping up and down stream to give accurate results or readings.

5. Pressure tappings should not be made on a noisy section of line, ie near bends, tee pieces or valves where turbulence may occur.

6. In a control system, a pressure tapping should keep time lag to the minimum.

7. Bonded strain gauge type transducers are best for marine use because they can withstand the normal range of vibration experienced.

8. Most control valves are mounted vertically to keep spindle and gland wear to a minimum.

9. It is an advantage to fit a diaphragm operated control valve, on assumption that the radiated heat more significant than the conducted heat.

10. If signal lines are wrongly routed, it can cause unstable systems.

11. Electric cables in high temperature areas can cause breakdown of insulation.

12. The type of piping used in machinery spaces for pneumatic control systems is copper piping.

13. Pneumatic piping should be run like electric cables in cable trays to provide support and prevent work hardening of the copper and cause problems in the future.

14. When laying out a control system, ease of maintenance should be considered for when future problems arise.

15. The hazard that is reduced when fitting receiver type gauges instead of direct reading gauges in control panels is that direct reading gauges could cause a fault or fire risk if they are to leak.

16. After the installation has been inspected for damage, a simple functional or calibration check should be carried out to ascertain that the accuracy of the individual instruments is still within the design limits.

17. A system cold run trial is vital because negligence at this stage could cause serious damage to a costly piece of plant.

18. The instrument calibration accuracy obtained by manufacturer's tests are plus or minus half percent full scale deflection or better.

19. It is absolutely essential that pneumatic control equipment is supplied with clean, dry air.

20. The usual maintenance with equipment with solid state modular construction is repair by replacement.

21. Poor performance is recognised by experience and by taking pen recordings and forming trends.

22. Before fault finding, the maker's instructions should be checked.

23. In fault finding, adjustments made blindly hoping the fault it will clear, can make it worse.

24. It is vital to realise the full benefits of automisation and that survey requirements are to be met without undue difficulty.

25. An increase in testing can increase the degree of human error and interference and disturbance to the system.

26. The length of time that a typical maintenance schedule should be based upon is that systems should be tested on a regular predetermined interval.

27. Good documentation of regular testing gives surveyors from any classification society proof that regular testing is undertaken and gives them confidence that the control installation is being maintained in good condition.

Engineering Pocket Book – Answers

13 DIESEL ENGINE THEORY & STRUCTURE

13.1 Answers – Engine Types

1. The two main types of diesel engine are two-stroke and four-stroke.

2. The type that tolerates low quality fuel best is two-stroke.

3. The type that dominates the ocean-going propulsion market is the two-stroke, slow speed type of engine.

4. The ratio of a long stroke, slow speed engine can be up to 3.8:1.

5. Medium speed engines are widely used in the ferry, RoRo, offshore market and those limited to size.

6. The different types of turbocharge system are pulse and constant pressure.

7. The type of turbocharge system that gives the best overall efficiency is the constant pressure system.

8. The original diesel engine operated on the principle of a cycle in which heat was added at constant pressure. This was achieved by the blast injection principle.

9. The theoretical cycle that modern diesel engines work on is known as the dual or mixed cycle.

10. The three reasons why the actual work done diagram differs from the ideal work diagram are:

    - The manner at which and rate of which heat is added to the compressed air (heat release rate) is a complex function of the hydraulics of fuel injection equipment and the characteristics of the operating mechanism, of the way spray is atomised and distributed, air movement, quality of fuel

- the compression and expansion strokes are not truly adiabatic. Heat is lost through cylinder walls
- the exhaust and suction strokes on a four-stroke cycle (and the appropriate phase of a two-stroke) create pressure differentials which the crankshaft feels as 'pumping work'.

11. A draw card gives an approximation to a crank angle based diagram.

12. The four operations within a diesel engine cylinder are: compression of a charge of air, injection of fuel which then ignites, expansion of hot gases formed during combustion and expulsion of the used gas to exhaust.

13. The engine stroke is measured as the full distance the piston moves between each end of its travel.

14. Engine timing refers to the relative time or position of the crank at which each operation during the cycle is commenced and completed.

15. TDC means top dead centre, where the crank is at a certain position, ie top.

16. BDC means bottom dead centre, where the crank is at the bottom.

17. That scavenging means air is entering the cylinder, expelling exhaust gas and recharging it for the next combustion.

13.2 Answers – Engine Construction

1. The main bearings in direct drive engines are aligned with the propeller shafting.

2. The material used for chocks instead of cast iron or steel is an approved non-shrink epoxy resin chocking material.

3&4. Holding down bolts are secured to tank tops by bolts or studs which are secured into the tank top and a nut fitted and locked underneath. Seals must be fitted to ensure the integrity of the watertight tank tops. Studs will normally have a clearance in the bedplate to allow for thermal expansion of the engine.

5. Holding down bolts are tightened hydraulically in a sequence to reach correct tension in the studs and compression in the chocks.

6. Chocks are hammer tested at regular intervals and additionally after heavy weather or damage.

7. The main bearings and crankshaft are supported by the bedplate.

8. The casting in the centre of the transverse web supports the main bearing and also holes for the tie-bolts.

9. The overall height of an engine can be limited by having the centre oil pan or sump lowered and so will require a recess in the ship's structure.

10. The type of welds used in bedplate construction are to be of a very high standard, carefully controlled and inspected, be stress relieved, shot blasted and tested for flaws. All plate edges must be prepared with double butt welds and complete penetration where possible.

11. Fatigue cracks start at points of high stress or sudden change in section.

12. Engine frames are known as 'A' frames because of their shape.

13. The bolts press up vertically within the frame, the pre-stressing of these maintains the frame in compression at all times.

14. The main strength member of a medium-speed engine is or may consist of a single frame or block which incorporates the crankcase, bedplate, frames and even the cylinder block.

15. The single structure is pre-stressed by bolts which pass through tubes constructed in the engine frames and entablature to the top of the cylinder block where locking nuts are hydraulically tensioned to pre-stress the structure, maintaining the cylinder block and frames in compression.

16. The tie bolt centres should be as close as possible to the crankshaft axis to reduce bending stress on the girders, to prevent unbalanced loads being transmitted to the welds.

13.3 Answers – The Crankshaft

1. The main bearings support the crankshaft at each crankshaft journal.

2. The properties that the crankshaft material must have are to be of high strength, have a long fatigue life and form good bearing surfaces. Carbon or low carbon alloy steels are used depending on size.

3. The process that improves fatigue resistance in medium-speed crankshafts is that the crankshaft is forged from one single billet of steel with its crystal grain structure in a longitudinal direction.

4. Oil gets to the big end bearings in a medium-speed engine by having holes drilled between journals and pins to transmit LO from the main bearings to the adjacent bottom ends.

5. A semi-built crankshaft is built by having each throw of the crank pin with two webs either forged or cast and then machined. Holes are bored in each web to allow them to be shrunk on the adjacent journals, which have also been forged and machined separately. The webs are then heated to the correct temperature and cooled slowly, shrinking them onto the journals to form a continuous length of crankshaft. Angles between units are corrected to correspond with firing order.

6. Dowels and keys are not used in semi-built crankshafts because they act as stress raisers and initiate fatigue cracks.

7. Overlap in a welded crankshaft permits large diameter journals and pins, reducing bearing pressures.

8. Two methods used to improve balance on crankshafts are to add external balance weights to the crank webs or drill holes in large diameter pins.

9. The type of bearings used in large cross-head engines are thick shelled white metal bearings.

10. In a direct drive engine, the thrust bearing is usually located within the engine. This makes use of engine bedplate and seatings to distribute propeller thrust to the ship structure. It also allows the bearing to form part of the engine LO system.

11. The type of bearing used as a thrust bearing is a single collar tilted pad type bearing which is usually fitted at the drive end of the shaft.

12. On slow-speed engines, the flywheel is usually only convenient for making crank angles and top centre positions.

13. Before turning the engine on turning gear, it must be ensured that all cylinder indicator cocks are open, as the motor used is usually of limited power.

14. The turning gear safety cut-out is to ensure the turning gear is removed before manoeuvring and to prevent the engine starting with the turning gear engaged.

13.4 Answers – Connecting Rods and Crossheads

1. The oil flow in the connecting rod of a crosshead is downward.

2. The connecting rod is kept as short as possible to limit the overall height of the engine.

3. The section of a medium-speed engine connecting rod is shaped like an 'I' shaped section girder to counter the transverse bending force subjected on it.

4&5. Engine guides which are fitted to crossheads are vertical sliding bearings and are made of white lined bearing material, lined with oil holes to lubricate the crosshead.

6. Excessive clearance between crosshead guides and the shoes causes noise, wear on the bearing end glands, uneven loads and fatigue.

7&8. The type of bearing used as the top end bearing is of the thin walled shell type (insert bearing) with a steel backing, supporting material of low friction such as white metal or aluminium tin alloy.

9. With the older type crosshead, the main concentration of load was on a simply supported beam which the connecting rod would locate on.

10. The equivalent of a crosshead on a trunk piston engine is a gudgeon pin in the piston skirt.

11. A non-return valve may be fitted to the foot of a connecting rod in a medium-speed engine to prevent back flow of oil under gravitational force at the bottom of each stroke.

12. Large diameter crank pins improve the rubbing and the rigidity of the bearing which has its length limited to the distance between each pair of webs on the crankshaft.

13. The bearing materials used in medium-speed bottom end bearings are thin wall bearing materials such as lead bronze, copper lead, white metal, aluminium, tin alloys, etc.

14. The harder metal bearing materials require or may require a thin, deposited overlay of very soft alloy to aid bedding in. They may require larger bearing clearances to allow for thermal expansion, have hardened polished journal surfaces and an increased supply of finely filtered oil.

15. The angled butts of medium-speed engine connecting rods are serrated to absorb side thrust and shear when the engine is running.

16. Three different arrangements of big end bearings on a crank pin in V engines are side by side on the crank pin, articulated (master and slave), fork and blade.

13.5 Answers – Pistons

1. Other components that require similar material properties to pistons are cylinder liners and covers.

2. The factors that the choice of materials for pistons depend upon are that it has to withstand high gas load and transmit the force from this to the piston rod. It must have a long fatigue life to survive fluctuating mechanical and thermal stresses.

3. The steel used for pistons in two-stroke engines is; chrome-molybdenum alloy steel, and to reduce wear and fretting of groove surfaces, they are chrome plated and ground.

4. To give pistons extra protection from high temperature corrosion, part of the piston crown surface may have a layer of protection alloy welded on.

5. The coolants used in pistons on slow, two-stroke engines is LO and water.

6. The coolant that has a higher thermal capacity is water.

7. The piston cooling system is kept running after 'finish with engines' in order to allow gradual reduction in temperature and consequent thermal stress.

8&10. The main purpose of a skirt on a large piston is when a uniflow scavenge is used. A short skirt may be fitted to act as a guide to stabilise the position of the piston with a liner. With two-stroke engines using loop or cross scavenge, long skirts are required to blank off scavenge and exhaust ports in the liner when the piston is above these.

9. Fitted to skirts of large pistons to assist running in are soft bronze bearing rings, which can be renewed if necessary.

11. The material used for the skirt in slow-speed engines is cast iron.

12. The purpose of the piston rod is to act as a strut and to transmit the heavy gas load downwards from the piston to the crosshead and running gear.

13. The piston rod is round in section for strength and rigidity.

14. The piston rod is attached to the piston in two-stroke crosshead engines by having the rod attached to the piston by means of flange at the top of the rod where bolts or studs are used.

15. The piston rod surface is treated to minimise friction and wear at the gland (stuffing box).

16. The top surface of a four-stroke engine piston may be recessed to allow clearance for exhaust and inlet valves which will be open on the overlap period of the engine cycle.

17. The length of a four-stroke piston skirt is dictated by the amount of side thrust it has to dissipate to the liner.

Engineering Pocket Book – Answers

18. Aluminium alloy pistons are not used when burning residual fuels because their strength, corrosion and abortion resistance is reduced at elevated temperatures.

19. The shapes that piston rings are machined to are circular or cam shape from which they expand to form a circle at working temperatures.

20. The three main piston ring joints are the butt (vertical), the scarfed (diagonal), lap or bayonet joint.

21. The two main purposes of piston rings are to convey heat to the cylinder liner and act as a gas seal between the liner and the piston.

22. The piston ring outward pressure is increased by gas pressure acting on the back of the ring.

23. When wear on piston rings becomes excessive, they should be discarded and new ones fitted.

24. Circumferential clearance of a piston ring is measured before fitting by inserting the ring into the least worn (lower) part of the liner bore.

25. The circumferential wear of a piston ring is measured by measuring reduction in the width of the ring section and by the increase in the butt clearance at the corresponding liner bore.

26. Coatings that are applied to piston rings may be chrome or plasma coatings.

27. The kind of liner not used with chrome plated or plasma rings is a chromium plated liner.

28. Some piston rings have barreled or grooved surfaces to assist build up of an oil wedge lubrication and accelerate bedding in and conformity.

29. Another name for an oil control ring is the scrapper ring.

30. The purpose of oil control rings or scapper rings is, on the upward stroke of the piston, to spread oil and on the downward stroke, scrape off any excess oil.

13.6 Answers – Cylinder Diaphragm and Piston Rod Gland

1. The cylinder diaphragm isolates the lower surface of the cylinder and scavenge box from the crankcase, and in doing so, prevents contamination of the crankcase LO from the residues of heavy fuel oil contamination.

2. The diaphragm, with regard to lubrication, allows the use of special high alkaline cylinder oils, while the main crankcase oil can be selected for better bearing lubrication and cooling properties.

3. The purpose of the upper half of the piston rod gland is to act as a seal for scavenge pressure and scrape off any residue dirt from the piston rod on its downward stroke. This contaminated oil residue should be conveyed to the sludge tank.

4. The purpose of the lower half of the piston rod gland is to act as oil control rings, to scrape off any excess crankcase oil during its upward stroke. This oil is returned via drains to the crankcase system. A void or vent space is left between the two sections and the drain should be inspected regularly to ascertain the efficiency of the gland.

5. The lack or incorrect maintenance of the piston rod gland can lead to contamination of the scavenge space with oil, loss of scavenge air, or contamination of crankcase oil and damage, with possible overheating of the piston rod, leading to danger of a hot spot within the crankcase and risk of an explosion.

13.7 Answers – Cylinder Liner

1. The base material of a cylinder liner is pearlite grey cast iron.

2. The alloying elements vanadium and titanium are added to enhance strength, wear and corrosion resistance.

3. To retain lubrication with chrome plated liners, the chrome must be deposited as a porous structure or else be etched.

4. A cylinder liner is secured by means of the upper end of the liner forming a flange of sufficient strength to support the liner. This flange is secured between the cylinder cover and the jacket or engine entablature.

5. The thickness of the liner can be reduced towards the lower end due to temperatures and pressures being less.

6. Inlet and exhaust port edges are shaped to direct the flow and reduce ring snagging.

7. The lower part of the cooling water space is sealed by silicone rubber rings.

8. The upper part of the cooling water space is sealed by the landing face of the liner flange on the entablature.

9. The term 'bore cooling' means a number of individually small holes are bored within the upper thickness of the liner close to the liner's surface to allow for better cooling where temperatures and pressures are greatest.

10. In large engines, the cylinder jacket consists of an individual iron casting, supported on the engine framing or entablature and bolted together to form a rigid block into which the cylinder liners are fitted.

Engineering Pocket Book – Answers

11. Cylinder liners require lubrication for piston rings and reduce liner friction and wear. The oil film also acts as a gas seal between the liner and the rings and as a corrosion inhibitor.

12. For heavy fuel oil burning crosshead engines, the TBN is 70.

13. The properties that cylinder oil for crosshead engines must have are to neutralise acids formed from combustion of sulphur in fuel, they must then retain the viscosity at high temperatures, resist oxidation and carbonisation, spread readily but yet cling to the liner surface. They must prevent scuffing and wear while keeping piston rings free from carbon build-up.

14. A cylinder oil must not be reused in a crosshead engine because alkaline agents will be expelled during use.

15. The cylinder oil is supplied to the lubrication quill by pressure pulses from positive displacement pumps, mechanical lubricators driven from the engine camshaft and regulated to deliver the required rate of oil.

16. Blowback of combustion gas is prevented in the cylinder lubrication system by each quill containing a non-return valve.

17. An accumulator in a cylinder oil system works by being fitted to the cylinder lubrication quill so that the correct quantity of oil will enter each time the cylinder pressure drops below that of the accumulator, ensuring oil does not enter the cylinder and be exposed to hot gases.

18. The cylinder oil is distributed over the length of the cylinder in a crosshead by the aid of gutters adjacent to the lubricator points and angled downwards to assist flow by gravity.

19. The supply of cylinder LO should be increased during running-in periods for new liners and piston rings.

20. With medium-speed engines, cylinder lubrication is usually achieved by oil splashing from the crankcase into the lower end of the liner. This would cause uneven excessive lubrication, so oil scrapper or control rings are fitted to the piston skirt to spread this circumfrentially and return excess oil to the crankcase.

21. Piston cooling oil is used in some medium-speed engines for cylinder lubrication by having oil bled from the piston cooling return and pass through small bore holes in the skirt and increase cylinder lubrication.

22. In medium-speed engines, insufficient lubrication is dangerous because of loss of oil seal to the piston rings which can lead to 'blow by' of hot gases causing local overheating, rapid breakdown of surfaces and seizure of pistons and possible risk of crankcase explosion.

23. Excessive lubrication can cause carbon deposits, piston rings slicking in grooves, allowing possible breakage or blow by. There will be fouling of exhaust system including turbochargers and contamination of scavenge spaces.

13.8 Answers – Cylinder Cover

1. The other name for a cylinder cover is a cylinder head.

2. The valves usually found in a slow-speed cylinder cover are an exhaust valve, air start valve, relief valve and indicator connection and fuel valves.

3. The cylinder cover cooling water comes from the jacket system.

4. Medium speed engine cylinder covers are made from castings of pearlite grey cast iron or, in some cases, cast steel. Deep sections are used to prevent bending under the peak pressures in high rated engines. The deep box like construction allows

air inlet and exhaust passages to be accommodated within the main casting.

5. The passages usually found in medium-speed cylinder covers are exhaust and air inlet passages.

13.9 Answers – Camshafts and Valve Gear

1. The valves and fuel pumps of an engine are operated by a cam follower which rises and falls as the cam rotates beneath it. The profile or shape of each cam is designed to give the correct timing, speed and height of lift to its corresponding follower.

2. Cams are made of steel with a hardened surface to the profile of the cam.

3. Some engines have a separate lubrication system to prevent any possibility of fuel contamination of the crankcase from leakage of a fuel pump.

4. The camshaft rotation must be accurately sychronised with the crankshaft and this timing must be checked periodically and after any adjustments or repairs have been carried out.

5. The cams on a four-stroke engine operate the exhaust and inlet valves, plus the fuel pump.

6. The speed that a slow-speed engine camshaft operates at is the same speed as the crankshaft.

7. The crankshaft is driven off the crankshaft.

8. The two types of camshaft drive are a train of gear wheels or a roller chain.

9. The camshaft drive used when the distance between the crankshaft and camshaft is large is the roller chain type drive.

10. The type of drive that gives a reduction in weight is the roller chain drive.

11. There is a high factor of safety in chains to prevent them stretching.

12. The three areas that wear occurs in the chain drive are between pins and bushings, between bushings and rollers and between rollers and sprocket wheel teeth.

13. Wear in the chain is measured by its extension.

14. The transverse movement in the chain drive should be roughly equal to one link pitch.

15. The approximate life of a camshaft chain is fifteen years, irrespective of wear.

16. At least one jockey wheel has an adjustable centre which allows the chain tension to be adjusted.

17. Usually the maximum adjustment of a chain is no more than 2% extension of its original length.

18. A pin link can only be riveted once and on no account a second time.

19. The timing is adjusted by slackening the camshaft flange and is adjusted to sychronise the cam timings with crankshaft position.

20. The timing is checked by either putting one engine unit at TDC or from corresponding marks on the flywheel.

21. Excessive tension in a chain will cause high load levels and possible damage.

22. The types of valves used for exhaust valves are of the poppet type with their valve lid and spindle of inverted mushroom

Engineering Pocket Book – Answers

shape, arranged to open inwards in order to maintain positive closing under pressure in the cylinder and ensure non-return action.

23. In four-stroke engines, the valves are actuated mechanically from the camshaft. Rotation of the cam peak raises the follower and push rod to operate a pivoted rocker lever, the other end of which depresses the valve spindle through a tappet and causes the valve to open.

24. The material used to make exhaust valves is a heat resistant alloy steel with stellite welded to the valve landing surface.

25. Stellite is a very hard alloy of cobalt, chromium and tungsten, with some iron and carbon. It resists impact, wear and corrosion at high temperatures.

26. The underside of some large exhaust valves are protected by a layer of sintered inconel or nimonic.

27. Exhaust valves are cooled by having the valve cage and seat circulated with fresh water.

28. The rotation of valves ensures and assists in removal of deposits between seat and landing face. It will also ensure even temperature and wear around the valve.

29. Tappet clearances are essential to allow for thermal expansion of the valve spindle at working temperatures.

30. The tappet clearance will tend to increase with of the seat during use.

31. An excessive tappet clearance will cause the valve to open late and close early in the cycle and will reduce maximum lift. It will also cause noise and eventually damage from the impact of working surfaces.

32&33. Insufficient tappet clearance will cause the valve to open early and close late. In extreme cases, it may prevent the valve from closing completely as it expands or beds in. This will cause hot gases to blow past the valve faces, causing valve burning, low compression, etc.

34. The hydraulic exhaust valve system replaces the push rod, rocker lever and tappet in mechanically actuated valves.

35. The pressure in the air spring is maintained by constantly being charged via a reducing valve from the air start system.

36. Air inlet valves are the same size as exhaust valves.

13.10 Answers – Turbocharge Systems

1. Engines are turbo-charged to increase power output and efficiency with only relatively moderate increase in size, weight and initial cost.

2. The charge air cooler improves the engine thermal efficiency by cooling the air at constant pressure, increasing its density before suppling it for compression in the engine cylinders. The mass of air per cycle can now be increased and the quantity of fuel injected can be raised to give a corresponding increase in engine output and thermal efficiency.

3. A turbocharger is said to be matched to an engine when at any given speed, the exhaust gas energy must cause the turbine to run at a stable speed at which the compressor will supply the correct mass of air to the engine.

4. The two operating systems for turbochargers are pulse and constant pressure.

5. The system that gives a rapid build-up of turbine speed when starting or manoeuvring is the pulse system.

Engineering Pocket Book – Answers

6. The system that gives high thermal efficiency is the constant pressure system.

7. A divided exhaust system reduces turbine efficiency.

8. The two parts of a turbocharger are the turbine side and the compressor side.

9. Turbine blades are attached to the turbine disc by a 'fir tree shaped root' design and they have free room to expand when heated. Binding wires are fitted to the blades to reduce vibrations.

10. The material used for turbine blades and nozzles is a heat resisting steel or nickel alloy.

11. The material used for the air impeller and inducer is aluminium alloy.

12. Fitted at the air inlet are filters which can be removed for cleaning and are also fitted with insulation to reduce noise.

13. The four parts that are changed together in a turbocharger to ensure matching are the turbine nozzle ring, air diffuser, impeller and inducer.

14. The type of seals fitted to turbocharger shafts are labyrinth type seals.

15. The seals prevent possible oil leakage into the turbine or compressor, or exhaust gas into the corresponding bearing oil.

16. The thrust bearing is fitted at the compressor end of the shaft, allowing the turbine end free to accommodate thermal expansion of the shaft.

17. The usual life of a roller bearing is 8000 hrs.

18. Updraught from the exhaust may cause the turbocharger to rotate while the engine is out of service.

19. The turbine bearing is cooled in an 'uncooled turbocharger'.

20. The uncooled turbocharger increases the thermal efficiency of a ship or plant by allowing more efficient use of the waste heat boiler.

21. Lower air compression temperatures are preferred as a charge air cooler is fitted to lower the air temperature between the turbocharger and engine air inlet manifold. Causing increased air density at lower induction temperatures and the lower compression temperature reduces stress on piston rings, piston and liner.

22. The air velocity is increased after the charge air cooler by having the outlet convergent to restore air velocity.

23. At the charge air condensate drain, the amount of moisture being removed from the air can be indicated or if a cooling water leak has occurred.

24. The air seal is maintained at the free end of the charge air cooler by means of a rubber joint ring.

25. The maximum air discharge temperature from a charge air cooler should be 55°C.

26. Too low an air temperature from the charge air cooler can cause thermal shock when in contact with hot liners and pistons.

27. Two-stage turbocharging gives higher charge air pressure and mass, increasing efficiency.

28. Charge air temperature should be kept above 20 – 25°C to prevent undercooling.

Engineering Pocket Book – Answers

14 FUEL SYSTEMS

14.1 Answers – Fuel Oil

1. The type of fuel used in modern marine diesel engines is heavy oil.

2. The three major problems associated with modern diesel fuels are:

   - Storage and handling
   - combustion quality and burn ability
   - contaminants.

3. The reason for build-up of sludge in modern marine fuels is because heavy fuels are blended from a cracked heavy residual, using lighter cutter stock, resulting in a problem of incompatibility.

4. To reduce the formation of sludge, a detergent type chemical additive is used.

5. Water in the fuel system can seriously damage fuel injection equipment, cause poor combustion and increase liner wear.

6. Water is removed from bunker tanks by draining or sludging tanks. Water can also be removed by proper operation of separators and properly designed settling and service tanks.

7. The significant problem caused by poor or incompatible combustion is the fouling of fuel injectors, exhaust ports, passages and the turbocharger gas side.

8&9. Ignition delay; a steeper ignition pressure gradient. These factors increase engine fatigue, excessive thermal loading, increased exhaust emissions, and critical piston ring and liner wear.

10. Four qualities used to indicate a fuel's burn ability are:

- Conradson carbon residue
- asphaltenes
- cetane value
- carbon to hydrogen value.

11. To achieve atomisation, particularly at low loads, it is important to ensure proper use of separators, settling tanks and filters, and correct fuel viscosity must be maintained.

12. The major fuel content influencing high temperature corrosion is vanadium.

13. The minimum melting temperature of vanadium compound of combustion is 530°C.

14. The action of a form of severe high temperature corrosion is where mineral ash deposits form on a valve seat which, with constant pounding, causes dents leading to a small channel through which hot gases can pass.

15. The main defense against high temperature corrosion is to reduce the running temperatures of engine components.

16. Prevention of high temperature corrosion is achieved by using intensively cooled cylinder covers, liners and valves as well as rotator caps fitted to valves. Special corrosion resistant coatings such as stellite and plasma coating should be applied to valves.

17. The action of low temperature corrosion is caused by sulphur in fuel. In combustion, sulphur combines with oxygen to become sulphur dioxide, some of this combines to become sulphur trioxide, which when combined with moisture vapours (from cooling temperatures being too low) becomes sulphuric acid, resulting in corrosion of cylinder liners and piston rings.

18. The danger of raising temperatures to combat low temperature corrosion is that it may lead to high temperature corrosion.

19. The normal abrasive impurities found in fuel are ash and sediment compounds, solid metals such as sodium, nickel, vanadium, calcium and silica which can result in wear of fuel injection equipment, cylinder liners, piston rings and ring grooves.

20. The new contaminant being found in marine fuels is the metallic catalyst composed of very hard abrasive aluminia and silica compounds, which are a cause for much concern.

21. The source of this contaminant is the carry over of the catalyst refinery process and remains suspended in the residual bottom fuel for extended periods.

22. Viscosity of a fuel was once considered the best guide to fuel quality.

23. The viscosity of a fuel is its resistance to flow and is a measure of the work done in moving a given mass of fuel. Viscosity is usually measured in Seconds Redwood or Degrees Engler from measurment using standard apparatus in which a given quantity of fuel is run through a standard orifice at a given temperature.

24. In the handling of modern marine fuels, it is necessary to keep bunkers from different origins separated they are not usually compatible and sludges can form.

25. When giving a viscosity, a temperature must be stated.

26. A cetane number indicates in a fuel the shorter the time between fuel injection and rapid rise in pressure.

27. Another name for the Conradson carbon value is the coke value and is a measure of the percentage carbon residue after evaporation of the fuel in a closed space under control.

28. The ash content in fuel is a measure of the inorganic impurities in the fuel. These are typically sand, nickel, aluminium, silicon and vanadium.

29. High sulphur levels in fuel are dangerous because of acid formation.

30. The water content in fuel is determined by centrifuging or distillation.

31. The pour point of a fuel is the lowest temperature at which an oil remains fluid and is important to know for onboard handling properties.

32. The flash point of a fuel is the lowest temperature at which an oil gives off combustible vapour, or the point at which air/oil vapour mixture can be ignited by a flame or spark.

33. The S.G of a fuel is used to calculate the quantity of fuel by weight in a tank of given dimensions.

14.2 Answers – Supply of Fuel Oil

1. The density of fuel must be given when bunkering because the consignment will be measured by weight. It will also be necessary to know this to adjust centrifuges to give efficient purification.

2. The viscosity of fuel is used to calculate the temperatures at which the fuel is treated and injected into the engine.

3. In the settling tanks, the fuel is heated to allow water and sludge to separate by gravity and be drained off.

4. The recommended treatment for fuels used in large slow-speed engines requires two centrifuges of adequate capacity, operating in series. The first acts as a purifier removing water, the second as a clarifier, removing solid particles.

5. The maximum temperature of fuel during treatment is 98°C.

6. To avoid fouling of fuel lines, a temperature of 150°C should not be exceeded.

7. The mesh of the fine filter in a fuel line is of stainless steel and able to filter out particles of up to 50 microns (0.05 mm) or less in smaller engines.

8. In the fuel system, various safety devices must be included with alarms, such as loss of oil pressure, low tank level, high fuel temperature, low fuel temperature and viscotherm failure, etc.

9. The maximum fuel oil storage temperature is 50°C or may not, if this is lower, exceed 20°C below the flash point.

10. To achieve correct atomisation of the fuel oil, it is necessary to reduce the high viscosity of the oil to a value at which correct atomisation occurs in fuel injectors.

11. The viscosity of fuel for combustion is between 10 – 15 centistokes at 50°C, that is 50 – 70 seconds redwood at 100°F.

12. A general indication of good combustion is a good power produced and exhaust temperatures normal for the fuel setting.

13. Four parameters for good combustion are viscosity, atomisation, penetration and turbulence.

14. The fuel droplet size depends on the size of the holes in the nozzle and the pressure difference between pump and discharge.

Engineering Pocket Book – Answers

15. Penetration with regard to combustion is the distance the oil droplets will travel into the combustion space before mixing and igniting.

16. The three conditions that penetration is dependent on are the droplet size, velocity leaving the injector and the conditions within the combustion chamber.

17. The spray pattern of the fuel is determined by the number of atomiser holes and their position.

18. Swirl to the air in the cylinder is imparted by the air during its entry at the scavenge ports.

19. Turbulence is important to combustion as it will improve mixing of fuel and air for effective and rapid combustion.

20. The term used to describe combustion in diesel engines is 'compression ignition'.

21&22. During the first phase of combustion, the atomised oil droplets emitted from the injector nozzle into the combustion space at the start of injection will evaporate and mix with hot compressed air and some chemical changes will take place. The mixture will reach an ignitable condition and spontaneous combustion will commence. The time elapsed during this phase is termed the ignition delay or lag.

23. During the second phase, the ignition and start of combustion will occur and set up a flame front which will accelerate through the chamber, enveloping and burning all the other droplets present, causing a very rapid generation of heat with corresponding rise in pressure and temperature.

24. If the injector continued to inject during the ignition delay and there was sufficient quantity, the rapid combustion and pressure rise will be quite violent, causing detonation, the shock loading creating a noise termed 'diesel knock'.

25. Ignition quality of fuel is a term used to denote its ignition delay.

26. The usual measurement of ignition delay is the cetane number.

27. Slow-speed engines can operate down to a cetane number of 24.

28. The lowest cetane number a medium-speed engine can operate down to is 34.

29. The ignition quality is important when starting an engine or when operating at reduced power for extended periods.

30. At low speeds, the increase in ignition lag can cause knocking or rough running of the engine.

31. Variable ignition timing (VIT) can be used to advance the start of injection, allowing for longer delay but maintaining ignition timing and the same peak pressure.

32. The design of the profile of the fuel pump plunger in a medium-speed engine allows for VIT to be built in.

33. The different methods of pilot injection are either two separate injectors are fitted or one injector for each unit which is designed to inject two charges to each unit.

34. The most notable feature of an engine with electronic ignition is the absence of a camshaft.

14.3 Answers – Fuel Pumps

1. The fuel pump plunger is operated by the cam follower operated from the camshaft.

2. Fuel delivery starts on a jerk type fuel pump on a fixed point on the upward stroke of the plunger, when the top edge of the plunger blanks off the suction port.

3. Delivery closes on the up stroke when the curved surface of the helix or scroll machined in the plunger uncovers the suction port. This allows fuel pressure above the plunger to fall to the suction pressure through a vertical slot or hole.

4. The quantity of fuel in a jerk type fuel pump is regulated by the vertical length of the helix when it is in line with the suction port. This setting is altered by rotating the plunger.

5. Injection timing is controlled by the relative angular position of the cam peak to the camshaft.

6. The two methods of varying the point of injection are to move the cam with respect to the camshaft or by raising or lowering the plunger with respect to the follower.

7. The plunger is lubricated by leakage of fuel in the small clearance between the plunger and barrel.

8. The delivery in a valve timed fuel pump is controlled by suction and spill valves.

9. Timing is altered by rotating eccentric pivots of the levers for the push rods.

10. VIT is achieved with this type of pump by having links which super-impose control on the suction valve pivot, giving adjustment of the start of injection according to fuel quality.

14.4 Answers – Fuel Injectors

1. The two basic parts of a fuel injector are the nozzle and the nozzle holder or body.

2. These two parts are joined by a nozzle nut.

3. The needle valve opens when the fuel pressure acting on the needle valves tapered face exerts sufficient force to overcome the spring pressure. The fuel then flows into the lower chamber and is forced out through a series of tiny holes. Once the injector fuel pump has ceased delivering the high pressure fuel supply, the needle valve will shut quickly under spring the compression force.

4. When the injectors are operating on heavy oil, a fuel circulating system is necessary and this is usually arranged within the injector.

5. The alignment of the injector oil passages is ensured by a dowel pin.

6. Prior to sailing, fuel injectors should be primed.

7. Defects that are found with fuel injectors are choking due to contaminants in the fuel, or carbon building up at the atomizer holes and leaking needle valves which may cause secondary burning of oil and reduce the combustion efficiency.

8. Exposed fuel pipes should be enclosed in a double skin tube, the outer skin will collect any oil leakage and return it to a safe place.

9. If the injection tip temperature is too high, carbon may form and impede the spray and corrosion may also occur.

10. The design of modern two-stroke diesel engines ensures the injectors are adequately cooled by transfer of heat to the surrounding bore cooled cylinder covers.

Engineering Pocket Book – Answers

11. The ideal position for a fuel injector is in the centre of the cylinder cover, allowing symmetrical, conical spray pattern in the combustion chamber.

12. When running at low power for long periods, combustion may be insufficient leading to fouling or wear.

15 COMBUSTION

15.1 Answers – Fuel and Fuel Burning (Diesel Engines)

1. The importance is that correct viscosity (which is defined as the internal resistance of flow between successive layers of the fuel) is required for pumping, atomising and burning of fuel.

2. Flash point is important as it is essential to know the lowest temperature at which fuel vapour will ignite when a flame is applied. It is also important to know for the storage of fuels.

3. Ignition point is important to know as it is the lowest temperature causing a fuel air mixture to ignite. The value varies considerably with grade of fuel, strength of mixture and the pressure.

4. The viscosity for diesel and heavy fuel oil are 5 centiskokes at 50°C and 350 centiskokes at 50°C respectively.

5. The importance of the cetane number is to know how quickly the fuel ignites after injection. A higher cetane number means good ignition quality.

6. The problem caused by high sulphur content in fuel is that sulphur burns to become SO_2 and SO_3. The latter combines with water to form sulphuric acid which is highly corrosive.

7. The problems caused by high vanadium content in fuel is that vanadium during combustion mixes with sulphur and sodium and they are converted into oxides which form chemically compounded ashes. These ash particles are melted and if they cool and solidify before striking any surface, they pass out of the engine as dust. But if they strike a surface having a lower surface temperature, then they will adhere, causing a build-up of heavy layers of slag

Engineering Pocket Book – Answers

as well as causing corrosion. The melting point of this ash is as low as 530°C.

8. Vanadium is completely soluble in oil but insoluble in water and so cannot be removed by water washing.

9. Sketch of a draw and power card for a two-stroke engine with a leaking fuel injector:

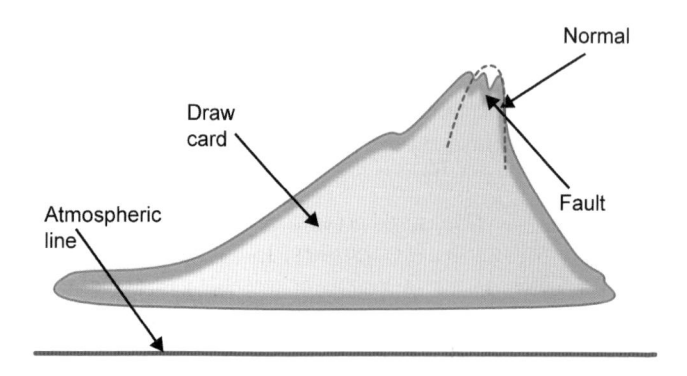

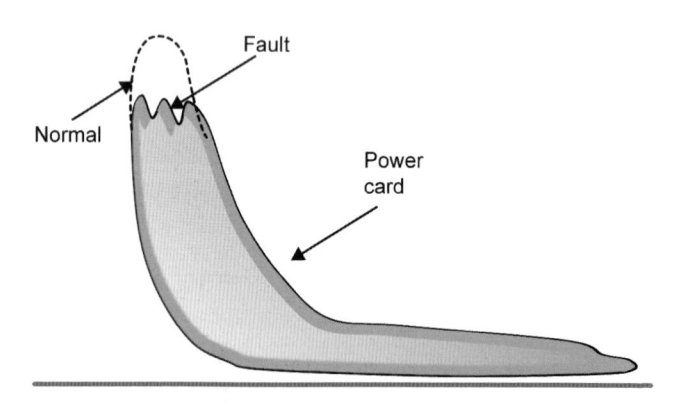

10. The consequences of a leaking injector are loss of power, high exhaust temperature and smoking exhaust. There may also be a knock or pressure wave in the injection system.

11. Sketch of a power card for after-burning.

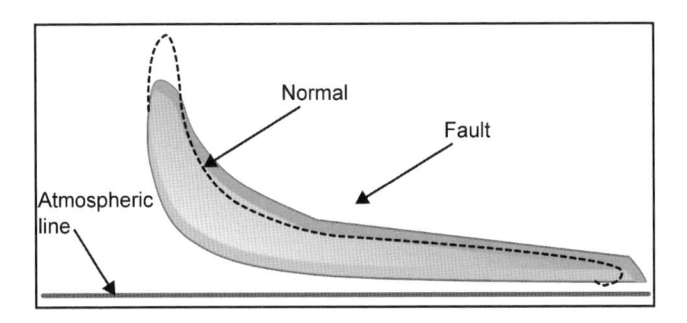

(Diagram showing a power card with Atmospheric line, Normal curve, and Fault curve labelled)

12. After-burning might occur because of slow or late combustion of fuel during the expansion stroke of the piston and is shown by a rise in exhaust temperatures and high pressure, with burning fuel and carbon passing to the exhaust. This may burn exhaust valves and foul the exhaust system with risk of the turbo-charger surging or up-take fires.

13. Power balancing is accomplished on medium-speed diesel engines as the power produced is related to quantity of fuel injected and balancing is carried out by small adjustments to individual fuel pump controls.

14. Compression diagrams are taken on slow-speed diesel engines with the indicator in phase with the engine's piston movement, but with the fuel shut off to the particular unit.

15. The consequences of an engine operating in an unbalanced condition are:

- Bearings and running gear will be overloaded, this may cause overheating and bearing failure

- overloading cylinders may cause piston blow by with corresponding dangers of overheated and seized pistons, also risk of scavenge fires
- unbalance will also set up vibrations which, if allowed for prolonged periods, will cause fatigue. This may lead to fatigue cracking of metal in bearings, fracture of bearing studs or bolts, cracks in the crankshaft or slackening or failure of holding down bolts.

16 LUBRICATION

16.1 Answers – The lubricating oil system

1. Items duplicated in a lubricating system are pumps, strainers and fine filters.

2. The lub oil pump draws from the lub oil drain tank via suction strainers, and is clear of any point to avoid picking up any water or sludge which may have settled.

3. The pump discharges at pressure through the oil cooler, the oil then passes through fine filters to the engine.

4. After the filters, the oil is distributed to all bearings, piston cooling, sprayers and exhaust valve actuators.

5. Drain returns are kept clear of the pump suction.

6. Drain returns must be submerged to prevent and reduce aeration and to make a safe seal.

7. The drain tank is generally situated into the ship's double bottom below the main engine crankcase.

8. There is a cofferdam around the drain tank to prevent any contamination from leakages.

9. The level gauge is situated centrally to reduce reading fluctuations.

10. The drain tank interior surfaces are coated to prevent rusting due to condensation.

11. A bypass centrifuge system is installed to improve the condition of the drain tanks contents.

12. The oil contaminant that should be eliminated or kept to a minimum is water content.

Engineering Pocket Book – Answers

13. The amount of oil used in cooling slow-speed engines is roughly twice as much as medium speed.

14. The range of pressure for slow-speed lubricating systems is 2 – 4 bar.

16.2 Answers – Lubricating Oils (1)

1. The base stocks of lub oils are obtained from fractional distillation of crude oil in a vacuum distillation plant.

2. Compound lub oils are non-mineral animal or vegetable (from 5 – 25%) added to a mineral oil to produce a compound oil.

3. Compound oils are used in the presence of water and steam because they tend to form a stable emulsion which adheres strongly to metal surfaces.

4. When sulpurised fatty oils gain an extreme pressure property.

5. In feed systems, corrosion, can be caused by fatty acid lubrication.

6. The total base number (TBN) is an indication of the quantity of alkali, ie base which is available in lub oil to neutralise acids.

7. The acidity of an oil must be monitored to avoid machinery damage.

8. Demulsibility refers to the ability of an oil to mix with water and then release water in a centrifuge.

9. Corrosion inhibition relates to the oil's ability to protect a surface when water is present in oil.

10. The condition known as scuffing is the breakdown of oil film on a surface causing instantaneous microscopic tack welding of a surface.

11. To maintain oil film under severe load conditions, special additives such as molybdenum disulphide (moly slip) are often used.

12. Lubricants used to prevent scuffing are extreme pressure lubricants.

13. Emulsification is associated with precipitation of sludge at an increasing rate.

14. With regard to emulsions, oxidation produces fatty acids which can lead to severe damage to machinery components.

15. Oxidation and corrosion products plus contamination products lead to deposit forming.

16. On cool surfaces, deposits form sludge of a soft nature.

17. Additives increase the life of lub oil by giving the oil properties it doesn't have, replacing desirable properties that may have been removed during refining and improving those naturally found in oil.

18. A doubling of oxidation rate is caused by every 7°C rise in temperature.

19. Oxidation in oil reduces the life of oil, plus viscosity usually increases.

20. Foam in oil prevents the formation of a good hydrodynamic layer of lubricant between the surfaces in a bearing and reducing the load carrying capacity.

21. A corrosion inhibitor is an alkaline additive used to neutralise acidity formed in the oil and, in the case of cylinder lub oils for diesel engines, to neutralise sulphuric acids formed from fuel combustion.

22. A detergent in an oil keeps metal surfaces clean.

23. Dispersants in an oil are used to stick to possible deposit making products and keep them in fine suspension by preventing small particles forming layers.

24. A dispersant is more effective than a detergent when at low temperatures.

25. A pour point dispersant works by adding a coat to wax crystals as they form when temperature is reduced preventing the formation of larger crystals.

26. The condition in a system that can lead to oil foaming is when air is entrained in to the oil, this could be due to low supply head or return lines not running full.

27. A viscosity index improver is added to help maintain viscosity of the oil as near constant with temperature variations.

28. The action of oiliness and extreme pressure additives are to reduce friction and wear. These additives would be important when running in gearing.

16.3 Answers – Lubrication Fundamentals

1. The functions of a lubricant are to reduce friction and wear, keep metal surfaces clean by carrying away possible deposits, provide a seal to keep dirt out and also to carry away heat generated in bearings and gears, etc.

2. The friction force with a slight trace of lubricant will change dramatically.

3. The two most important properties of a lubricant are oiliness and viscosity.

4. To improve the oiliness of an oil film, colloidal suspension graphite is added.

5. When there is no material contact, friction is determined by viscosity.

6. In the crankcase, boundary lubrication is said to exist in some top end bearings and guides.

7. Hydrodynamic lubrication is a condition where the bearing surfaces are completely separated by an oil film or layer.

8. The maximum oil pressure in a bearing is at the inlet (initial point in direction of rotation).

9. Four factors affecting hydrodynamic lubrication are viscosity of the lubricant, relative speed of the surfaces, bearing clearance and pressure, ie bearing load per unit area.

10. Insufficient circulation of lub oil means high temperature because the heat being generated by the bearing isn't being removed.

11. If rotational speed is kept constant, relative speed can be increased by increasing a journal or crank pin in size.

12. High bearing pressure can lead to boundary lubrication.

13. The kind of lubrication found in a bearing when it starts to move is boundary lubrication.

14. As the shaft speeds up, the oil in the journal bearing is dragged behind the pin until the oil film breaks through and separates the surface.

15. In a Michell bearing, the bearing surface consists of divided kidney shaped pads extending part or all the way round the surface. This principle is used in tunnel or shaft bearings.

The pads are prevented from moving circumfrentially but are free to tilt and incline in the direction of rotation. Such tilt allows a self-adjusting oil film wedge giving full film lubrication.

16. Pressure in order of 30 bar can be found in Michell bearings.

16.4 Answers – Lubricating Oils (2)

1. The range of TBN that slow-speed engine crankcase oil has is 10 – 15 TBN.

2. To measure the condition of lub oil, a sample should be taken from a working system at regular intervals and sent to a laboratory for detailed analysis.

3. A blotter test indicates oxidation, carbon present and dispersive properties when compared to similar drops of fresh oil.

4. The problem of cylinder lub oil has increased by the use of low quality, high sulphur content, residual fuels.

5. The local surface temperature can be higher than the gas temperature at TDC due to poor combustion and flame impingement on the cylinder walls.

6. The range of surface temperature that may be found at the top of the piston ring area is 180 – 220°C.

7. The range of temperature found at the lower end of the cylinder jacket may vary with type of engine, ie cross scavenged or uniflow. Cross scavenged have higher temperatures than uniflow, but temperatures are around 90 – 120°C.

8. Thermal cracking of lub oil begins at around a temperature of 300 – 350°C.

Engineering Pocket Book – Answers

9. The six requirements of cylinder lubrication are that it must:

- Reduce sliding friction between rings and liner to minimum
- possess adequate viscosity at high working temperatures
- form an effective seal in conjunction with piston rings
- burn cleanly, leaving as little soft deposit as possible
- effectively prevent build-up of deposits in the ring zone
- effectively neutralise the corrosive effects of mineral acids formed during combustion of fuel oil.

10. The disadvantage of paraffin oils as cylinder oils is that they oxidise at high temperatures leaving hard carbon deposits.

11. Wear rates that are considered normal for cylinder liners in slow-speed engines are 0.1 mm/10000 hours or even, in many cases, 0.05 mm/10000 hours.

12. The first type of cylinder lub oil developed for residual fuel burning was the emulsion type cylinder oil.

13. The disadvantage of the dispersive type cylinder oil was that when it was stored for long periods, a small proportion of the additive came out of dispersion, forming sludge like deposits in service tanks and lubricators.

14. The neutralising agents used in modern single phase cylinder lub oils are complex compounds containing a high percentage of calcium carbonate.

15. The alkalinity of an oil is expressed as the Total Base Number (TBN).

16. The speed of the neutralising reaction is important because the more rapid the neutralising reaction, the less time available to attack the metal.

Engineering Pocket Book – Answers

17. Spreadability of an oil is the properties it has to spread and flow sideways as well as downwards.

18. The storage stability of an oil is important with regard to oil lines so it doesn't settle out to harden and block lines in the form of sludges.

19. The percentage rarely or seldom exceeds 0.05% fuel consumption.

20. Chromium plating is beneficial to wear rates as this alloy is resistant to attack by mineral acids.

21. The oil flow rate to bearings is greater than that needed for lubrication only because greater weight of alkaline additive is necessary to prevent corrosive wear.

22. The percentage of additives in cylinder lub oil can be up to 33%.

23. The contact pressure in modern bearings is around 176 kg/cm^2, which is around the limit of safety.

24. The drawback of increasing the viscosity of lub oil is poorer cooling in bearings and, more importantly, in oil cooled pistons.

25. To achieve low under crown cooling in pistons, the design has to be good to allow efficient cooling.

26. In piston cooling, two conditions that must be at a maximum are the heat transfer, maximum oil velocity and turbulence.

27. The type of cooling that the motion of the piston creates is the 'cocktail shaker' type cooling.

28. Oxidation of organic mineral oil forms a lacquer, gum, organic acids and carbonous compounds.

Engineering Pocket Book – Answers

29. The result of thermal cracking of oil is carbonous compounds.

30. Thermal cracking is prevented by not subjecting oil to such high temperatures, or at least to minimise the period of exposure.

31. Cold sludge in lub oil is formed under cold operating conditions. This is a complex mixture of solid or semi-solid oil detergent products from incomplete combustion of fuel and lub oil.

32. Sludge is more likely to be found in trunk piston engines than crosshead engines because crossheads have a diaphragm and piston gland to separate cylinder from crankcase.

33. Lub oil centrifuging is carried out at a temperature of around 82 – 85°C.

34. The corrosion inhibitor that does not work with salt water contaminated oil is an alkaline additive based inhibitor.

35. Oil additives must not be water soluble as they will be removed in the centrifuge.

36. Stable emulsions must be avoided, otherwise the mixture might thicken to a degree where lubrication is impaired.

37. The viscosity index that a good crankcase oil should have is 75 – 85.

38. To improve the load carrying capacity of an oil, organic oiliness agents are added.

39. The same oil is used because considerable admixture would occur if two separate oils were used.

40. Poor quality fuel means, in the operation of an engine, that there is greater deposit formation of incompletely burned constituents and therefore the higher the detergency requirements in lubricating oil.

41. The lubricating properties of an oil are determined by its number, ie SAE 20, SAE 30, SAE 40.

42. As the TBN increases, as well as the alkalinity of the oil increasing, the ash content also increases.

43. The load carrying additives that combat valve train wear and varnish formation are zinc based additives.

44. The TBN of an oil is used in sulphur burning medium speed engines is TBN 22 – 34.

45. The properties of oil in high-speed engines are good thermal stability and high temperature detergency/dispersing qualities.

46. The oils sometimes preferred in lifeboat engines or emergency generators are SAE 20 oils, because cold starting may be important, especially in an emergency.

16.5 Answers – Shipboard Lubricating Oil Tests

1. The only place a complete and accurate picture of the oil condition can be obtained is a laboratory.

2&3. The procedure for the alkalinity test is to place a drop of indicator solution on blotting paper followed by a drop oil sample placed at the centre of absorbed indicator. A colour change takes place in the area surrounding the oil spot. If it turns red, this shows acid is present, if it turns blue/green, this shows alkali present. If a yellow/green colour appears, this shows it is neutral.

4. To determine the condition from an oil spot on blotting paper, the shape, colour and distribution of colour of the spot gives an indication of the oil condition.

5. If there are contaminants at the centre of the oil spot, this shows dispersiveness is poor.

6. The procedure for a flow stick test (Mobil) requires two equal quantities of oil, fresh and used samples of the same grade of oil. Equal capacity reservoirs are filled with these oils and allowed to reach an equal temperature. Then the device is tilted from the horizontal and the oils flow down parallel channels. The flow stick has reference marks along its length for comparing the viscosity of the used oil with the fresh.

7. Reduction in viscosity of an oil could occur because of dilution of fuel.

8. An increase of oil viscosity could happen due to heavy carbon contamination or oxidation.

9. Large amounts of water are detected in oil by a simple settling test.

10. When microbial degradation occurs in oil, a sludge or slime is formed.

11. Oils that are prone to microbiological infection are oils with water in.

12. The maximum recommended water content for detergent oil is 0.02% water content.

13. As a precaution against microbiological degradation of an oil, jacket and piston cooling water should be treated with additives that do not feed microbes.

14. Biocide is added to oil to combat microbiological infection.

16.6 Answers – Grease

1. Grease is a semi-solid lubricant consisting of high viscosity mineral oil and metallic soap with filler.

2. Fillers used in grease are lead, graphite and molybdenum disulfide. Fillers enable grease to withstand shock and heavy loads.

3. The advantages of a grease are that it will stay in place, will lubricate, will act as a seal and is useful for inaccessible parts.

4. Sodium soap greases are suitable for high temperature and high speed use.

5. Sodium soap grease will emulsify.

6. The melting point of sodium soap grease is about 200°C.

16.7 Answers – Oil Drain Tank

1. In large engines, the oil drain tank is situated directly below the engine in the ship's double bottom.

2. The recommended width of the cofferdam surrounding the oil drain tank is 18 inches minimum.

3. The purpose of the cofferdam surrounding the oil drain tank is to prevent contamination and also as an inspection tank for leakage or corrosion.

4. The oil drain tank should be as deep as possible, to present the minimum surface to air, reducing oxidation, air entrainment and foaming.

5. The oil drain tank can be made deeper than the ship's double bottom by locating the tank alongside the engine. The lower part is built into the double bottom and the upper part extended above the main tank top.

6. The bottom of the drain tank is sloped and 'V' shaped to enable water and sludge solids to collect at the lower end, where they can readily be removed.

7. An inverted 'top hat' is fitted below the lowest point of the oil drain tank to facilitate removal of sludge/water by means of a hand pump.

8. Oil returns should be short and direct.

9. Piston cooling oil passes through sight glasses outside the engine.

10. The oil returns should preferably be at the after end of the tank, they should be extended well below the minimum level.

11&12. The oil pump suction in a drain should be at the end remote from the returns, which avoids short circuiting between the two, giving sufficient time for deaeration and settling out water and foreign matter.

13. The oil pump suction should be at least 100 mm above the tank bottom.

14. Internal baffle plates should be fitted to the oil drain tank, both to strengthen the tank and also reduce turbulence and promote settling out of foreign matter.

15. Drain holes are left at the bottom of the baffles to allow water and solids settling out to return to the lower end for removal.

16. With a full charge of oil in the tank, there should be an ullage of at least 15 – 20 cm.

17. Strengthening ribs should be fitted to the outside of the oil drain tank to present a smooth, unbroken internal surface. Internal ribs act as traps for sludge, solids, etc and promote internal corrosion.

16.8 Answers – Pre-Cleaning and Corrosion Protection

1. Before putting bulk oil tanks into a system, they should be shot blasted under construction and immediately painted with oil, water, acid and alkali resistant paint.

Engineering Pocket Book – Answers

2. When using protective paints, it is essential that good preparation has taken place.

3. If a layer of moisture or rust forms before applying the paint, it will prevent the paint bonding to the metal and, what is worse, corrosion will continue below the protective surface.

4. It is better not to use protective paint if proper surface preparation cannot be ensured.

5. In smaller engines with wet sumps, is a circulating pump usually fitted into the lub oil system.

6. The purpose of a hand pump fitted in the lub oil system of smaller engines is to circulate oil through bearings prior to starting.

7. The importance of cleaning an entire lub oil system after construction or erection is to flush or remove any dirt, rust or welding scale as they can cause serious damage to bearings, journals and working parts.

8. If paraffin is used for cleaning, it must be removed after use.

9. Grease and oil type preservatives should be removed by wiping down using rags soaked in clean oil.

10. Flushing of a system should be carried out for two days.

11. The normal temperature of the oil drain tank is 32 – 43°C.

12. The most convenient method of heating the flushing oil is by passing hot water or LP steam through the waterside of the oil cooler.

13. The heating of flushing oil is most important with cast iron components in lub oil as it tends to trap fine foundry sand in its pores and heating causes an expansion of metal which releases them.

14. The circulating of flushing oil can be stopped when no further solids are trapped in the temporary filters.

15. After removing the temporary filters, the next procedure in the flushing operation is to continue flushing for a further 24 hours with the engine turning.

16. The kind of oil used for flushing instead of special flushing oil can be a good quality clean SAE 30 crankcase oil.

17. The viscosity of an oil can be lowered by heating.

18. The same oil should not be used for several flushing operations because of contamination.

19. The increasing problem with lubricating oils onboard ship has been accidental mixing of different oils.

20. A multiple tank is not suitable for oil storage as leakage may occur between tanks, through dividing bulkheads.

21. A separate filling line is essential for each oil storage tank to prevent contamination from different oils.

22. Deck level flush filling lines are not recommended as such plugs are rarely watertight and serious contamination can occur from seawater finding its way to storage tanks.

16.9 Answers – Rotary Displacement Pumps

1. Rotary displacement pumps have a lower efficiency than reciprocating pumps due to large areas with running clearances being exposed to pressure differentials between the discharge and suction pressures.

2. Rotary displacement pumps are almost exclusively used for pumping oils because the lubricating qualities of the used oil reduce wear on moving parts to a practically negligible amount.

3. The two types of displacement pumps are gear pumps and screw type pumps.

4. The number of threads in a counter screw pump determine the pressure for which the pump is designed.

16.10 Answers – Coolers

1. The usual cooling medium in coolers is seawater.

2. In a tube cooler the oil is in contact with the outside of the tubes and shell of the cooler.

3. Tubes in a cooler are usually made from aluminium brass (76% copper, 22% zinc, 2% aluminium).

4. If the water boxes of coolers are not made of steel or coated, they should be fitted with sacrificial anodes.

5. Early tube failures in coolers may be caused by pollution in coastal waters or, in some cases turbulence.

6. Tube plates are secured in a cooler by having one end that is fixed and the other end that is free to move with expansion of the tubes.

7. To detect joint leakage without mixing the fluids, there is a special 'tell-tale' that will allow the fluid to escape without mixing with the other fluid.

8. The cooler shell is manufactured out of fabricated or cast iron.

9. The shell material is not critical as it is not in contact with seawater, so there are no corrosion problems.

10. In horizontal set coolers, seawater should enter at the bottom and leave at the top.

11. Vent cocks are fitted to coolers so that any air can be purged from them, as air in the system will encourage corrosion and reduce the cooling area, causing overheating.

Engineering Pocket Book – Answers

12. The recommended procedure for controlling the oil temperature through the cooler is by adjusting the seawater valve on the outlet from the cooler.

13. Controlling temperatures by flow increases the risk of corrosion.

14. Partial blockage in a tube cooler can cause the cooler to become ineffective and tends to lead to corrosion.

15. The kind of brush used for cleaning cooler tubes should be special soft brushes.

16. Chemical cleaning of tube coolers is recommended when hard deposits have accumulated in the tubes.

17. Before handling chemicals, contact should be avoided by wearing gloves, face shield and an apron and instructions must be followed.

18. After using a chemical agent, flushing is necessary.

19. Plate type heat exchangers were first developed for the milk industry where cleaning is essential every day.

20. The effect on heat transfer that the corrugation has on the plates of the cooler is that it gives a turbulent flow and increases the heat exchange area.

21. In smooth flow, a boundary layer acts as a heat barrier.

22. Tube cooler materials are not suitable for plate type coolers because turbulence can cause damage due to erosion in materials normally used in tube coolers.

23. The usual jointing material in plate type coolers is Nitrile rubber.

24. The joint between the plates is maintained by an adhesive such as Plibond, the plates are then clamped together.

25. For high temperature joints, compressed asbestos fibre is used.

26. Over tightening of a plate type cooler can cause damage to the chevron-corrugated plates.

27. The flow ports are arranged in a cooler plate at the fixed end.

28. To indicate leakage, in the double-jointed section of a plate cooler, a drain hole acts as a tell-tale for this section.

29. No extra space is required for dismantling this type of cooler.

30. The capacity of a plate cooler is altered by adding or removing plates in pairs.

31. Leak detection is more difficult in a plate type heat exchanger.

32. Lubricating oil coolers are usually of the tube type because of the pressure differential.

33. The major drawback of plate type coolers is the cost as there are a large number of joints on the cooler and the plates are expensive.

34. Cracks are detected in the plates of plate coolers by spraying dye penetrant and viewing under an ultra violet light to show up any defects.

35. The main benefits of titanium are that it is highly corrosion resistant, it is light and has good strength.

16.11 Answers – Oil Maintenance

1. The different types of contaminant in lubricating oil are water, aqueous mineral or organic acids, dirt, dust, rust, scale, wear particles from bearings, cylinders, gears, paint and jointing compounds.

2. Contaminants that are formed from incomplete combustion of fuel are carbonaceous compounds and sludge.

3. The five classes of filtration equipment are:

- Simple settling out of contaminants under static conditions
- centrifugal separators
- mechanical strainers and filters, coarse and fine
- absorbent and absorbable filters
- chemically active filters.

4. For a given throughput, the relationship between filter size and filter mesh size is that the finer the filter, the lower the throughput.

5. The largest particle size passed through a Vokes multi-element filter is 10 microns.

6. The elements in a Vokes multi-element filter are made from fine felt (throw away type).

7. Coarse lub oil filters are normally fitted to the suction side of the lub oil pump.

8. Fine lub oil filters are normally fitted at the discharge side of the lub oil pump.

9. Until recently, the majority of fine lub oil filters could not tolerate large quantities of water.

10. An efficient purifier can continuously handle up to 10% water contamination.

11. The maximum rated throughput of a purifier is not the same as the optimum efficiently throughput.

12. Filters are fitted in pairs to allow cleaning, so the system doesn't have to be shut down and there is always a filter on standby.

Engineering Pocket Book – Answers

13. A filter strainer can normally be cleaned with compressed air or by brushing. It should be cleaned as soon as it is out of the system, then re-assembled and left ready for use.

14. Usually, the smallest particles removed by a wire mesh type strainer is rarely below 125 microns.

15. The size of particles that can get through an Auto-Klean filter are particles down to 25 microns.

16. Pressure gauges are fitted before and after filters to give an indication of differential pressure across the filter and its condition.

17. The filtering mediums found in fine filters are natural or synthetic fibrous woollen felt or paper filtering mediums.

18. The type of filter that can be fitted in a bypass lubrication line is a centrifugal filter.

19. The basic type of marine oil centrifuge is the large bowl type, fitted with discs.

20. The force that acts on particles in a centrifuge is a radial outward centrifugal force.

21. Dirt particles move down the underside of the discs in a centrifuge against the oil flow by finding their way to the underside of the disc and enter a region of zero velocity. They can then move due to centrifugal force down the underside of the disc and eventually into the sludge space of the bowl.

22. The factors that effect the size of particle removed are:

- Velocity of the oil in the centrifuge
- disc spacing, diameter and inclination to the vertical
- speed of rotation of the bowl
- throughput.

Engineering Pocket Book – Answers

23. Centrifuge bowls are made up to a diameter of 0.6 m.

24. Continuous operation of a centrifuge is maintained over a long period of time by an ejection process which is timed to discharge sludge at regular intervals.

25. When overhauling centrifuges, it should be noted that some threads are left handed and that all parts should be handled with care as the centrifuge is a perfectly balanced piece of equipment, rotating at high speeds (5000 – 8000 rpm).

26. The best set up for the continuous bypass purification of lubricating oil is to take oil for the purifier from a point in the lubrication system where oil has passed through the system and had time to settle, and therefore should be at its dirtiest.

27. After engine shutdown, the purifier should be left running for twelve hours in order to minimise corrosion due to vapours condensing as the engine cools.

28. In a multi-engine installation, the minimum number of purifiers recommended for purification is one per engine.

29. Lubricating oil water contamination can combine with sulphur combustion products to form sulphuric acid.

30. Pre-washing lubricating oils before centrifuging is beneficial to remove acid in the oil formed from combustion.

31. The washing water should be 5°C above the heated oil.

32&33. Fine filtration is still necessary with good detergent oils because although detergent oils are able to carry fine solids in suspension, well below 1 micron, solid contaminants from induction air can be brought in which are many times the size that can be carried in suspension, so fine filters are still required.

34. As solid content increases in detergent oils, a point is reached when the very fine particles begin to agglomerate to form much bigger particles.

35. The recommended time for oil changes in medium/high and high-speed engines is 500 – 1000 hours. Large trunk piston engines, upwards of 15000 hours.

17 STARTING & MANOEUVRING SAFETY

17.1 Answers – Starting Air

1. The method used to start main propulsion engines is by the use of starting air, usually around 30 bar.

2. The opening of the air start valve is dictated by pilot air, which is supplied from the air start distributor.

3. In modern practice, air is introduced to the cylinder slightly before TDC, this allows air to accumulate in the clearance volume ready to force down the piston once it is over TDC. At the same time, another cylinder will be receiving air (because of overlap), this unit will be one in which the crank is well over TDC so that it generates an adequate turning moment to carry the above unit over TDC.

4. In indicator diagrams, the area under the curve represents the energy used in the starting operation.

5. The non-return valve should be fitted between the engine and the air receiver in an air start system.

6. A relief valve or bursting disc or cap should be fitted between the non-return valve and cylinder in an air start system.

7. Starting air is stored at a pressure of around 30 bar.

8. The two connections on the air start distributor are:

    - One for depressing each timing valve in the distributor from its free position to engage it with the timing cam
    - the second supplies air to the operating ports, from which it will pass through any timing valves that are open, to the corresponding cylinder valves, causing them to open.

Engineering Pocket Book – Answers

9. Timing valves are kept clear of the starting air cam by springs.

10. In multi-cylinder vee engines, a possible arrangement is to have the starting air valves in one bank of the cylinders.

11. A leaking air start is detected when the engine is running by checking the temperature of the adjacent pipe to the valve.

12. The capacity of air (in starts) that a receiver must store for a reversible engine is 12 starts.

13. The preferred method of fitting a receiver's fittings are on a valve manifold with one common connection to the shell.

14. The pressure gauge of a receiver must be in direct communication with the internal pressure, irrespective of other valves being opened or closed.

15. A drain valve fitted to an air receiver must be of sufficient size to prevent choking.

16. Air received from the compressor should be free of oil and moisture and not at excessive temperature.

17. Before opening a receiver's manhole door, it must be assured that any pressurised part of the system is isolated by locked valves, also two valve isolation is recommended. Internal pressure is completely discharged. The door opens inwards and if undue force is required, it should be confirmed that no pressure remains in the receiver.

18. During internal cleaning of an air receiver, care must be taken to avoid debris entering connections.

19. During internal inspection of the air receiver, careful inspection must be made for corrosion.

20. When applying protective coatings, all surfaces must be dry and clean and connections plugged. Protective clothing and goggles should be worn and a second person outside the manhole must keep watch in case the person inside is overcome by fumes.

21. There must be two starting air compressors, each capable of supplying all demands. One must have an independent drive for emergency use.

22. The maximum compressor delivery air temperature should not exceed 93°C.

23. Manual starting is attractive because it is both inexpensive and simple.

24. The type of drive used in gear drive starters is a Bendix drive.

25. The voltages of electrical starters are 12 and 24 volts DC.

26. Batteries are connected in series for starting.

27. The two types of battery used for electric starting are lead acid and nickel-cadmium alkaline batteries.

17.2 Answers – Manoeuvring/Direct Reversing Engines

1. To manoeuvre a ship in reverse the propeller thrust must be reversible by means of reversing the propeller drive or by altering the propeller pitch.

2. The timings in a four-stroke that need to be adjusted to run in reverse are fuel pumps, inlet valves and exhaust valves. To do this, a set of astern cams are fitted.

3. On four-stroke engines, to allow ahead and astern running, a separate set of astern cams are fitted, each cam is fitted adjacent to its corresponding ahead cam. The reversing procedure is carried out by moving the whole camshaft axially, which moves ahead cams clear of their followers which engage the astern cams.

Engineering Pocket Book – Answers

4. On large two-stroke engines, to run astern, fuel pump timing needs to be adjusted.

5. Gearboxes are fitted because the shaft speed of medium-speed diesels is not suitable where a low-speed propeller is required.

6. A clutch is a device to connect or separate a driving unit from the unit it drives.

7. With a unidirectional propulsion unit, two clutches are fitted, an ahead and an astern clutch.

8. The ratios of modern gearbox installations range from about 2:1 – 4:1.

9. The type of gearbox used in medium-speed drives is a single reduction and usually single helical.

10. The usual design for an internal gearbox clutch is the plate type clutch which consists of pressure plates and clutch plates arranged in a spider.

11. Pinions are attached to their shafts by keys or increasingly being fitted with the keyless oil injection method, on an appropriately tapered seat.

12. The kind of bearings found in early gearboxes were usually plain, and thrust of classical Michell pattern. But commonly now, roller bearings are used.

13. The controllable pitch propeller is attached to the tailshaft by means of a flange.

14. The oil reaches the operating mechanism of the controllable pitch propeller through the hollow shaft to the hub.

15. The duty of the early forms of inertia type governors were to act as overspeed trips. These governors are now largely obsolete.

Engineering Pocket Book – Answers

16. The fluctuation of the governor is called hunting.

17. Centifugal governors are not suitable for driving alternator engines because they cannot be made truly synchronous (constant speed).

18. The optimum speed of a centrifugal governor is 1500 rpm.

19. There is a tendency for wear to be only over a small area of the drive shaft because the operating range is small.

20. If the governor spring were to fail, the flyweights would be able to fall outwards and shut the engine down.

21. The governor spring is a trumpet or conical shape as this gives a more effective relationship between spring compression force and centrifugal force on the flyweights.

22. Droop is the fall in speed of the engine when load is applied.

23. The difference between coarse and fine droop is that when a small reduction in speed occurs as the load increases, the governor is said to have fine droop, coarse droop occurs when final running speed drops well below the desired value.

24. To indicate the load limit of an engine, limits of engine exhaust temperatures, jacket temperatures are taken from trial data, and most governors have built-in load limiting devices.

25. The governor should be situated as close as is practical to the engine fuel pumps, limiting the mass/inertia of the operating linkage.

26. If the oil in a governor is too thick, it will slow down the reaction time of the governor.

27. The two factors essential for the production of generated voltage are rotational speed and magnetic flux.

28. The usual droop found with marine governors is 5%.

Engineering Pocket Book – Answers

29. To provide droop in the AVR, a quadrative current compensation circuit is used (QCC), consisting of a current transformer (CT) and resistor.

30. Overspeed trips are fitted as it is possible that the engine governor may fail or not act quickly enough in an emergency situation to prevent damage to the engine. They should act independently of the main speed governor and, in the event of overspeeding, will immediately cut off power to the engine, either by raising the fuel pump plunger followers clear of the cams or by opening fuel suction valves, either of which will shut off fuel and power.

17.3 Answers – Safety Systems (Crankcase)

1. The size of a crankcase explosion is governed by available volume of the explosive vapour and it is this that makes large slow-speed main engine explosions potentially devastating.

2. The normal operating atmosphere in the crankcase will contain oil droplets formed by lub oil splashing from the bearings on to surfaces. This mixture will not readily burn or explode.

3. The most common cause of lowered lub oil flash point is contamination with fuel oil.

4. The minimum temperature considered for a hot spot is 360°C.

5. The range of air/oil mist ratio that is most dangerous is the middle of the range, 5 – 7% oil fuel vapour in air.

6. The secondary explosion is the most dangerous.

7. There should be no cross connections between crankcases to prevent a chain reaction in the event of an explosion in one engine.

8. A hot spot in the crankcase can be indicated by irregular running, engine noise, increase in temperatures and by the presence of white oil mist.

9&10. The best course of action when a hot spot is suspected is to stop the engine (check with bridge first), increase lub oil flow, engage turning gear and turn with indicator cocks open to prevent seizure of overheated parts.

11. The crankcase should be allowed to cool for at least 30 minutes before entering.

12. To ensure that crankcase explosions never occur, good regular maintenance of the engine, avoidance of overloading and a provision of adequate lubrication should be ensured.

13. The principle behind oil mist detection is to monitor samples of the air and vapour mixture drawn continuously from the crankcase of a diesel engine. Such a device will detect the presence of oil mist at concentrations well below the level at which explosions may occur, giving an early warning in time to allow action to slow and prevent either serious damage or an explosion.

14. Oil mist sampling pipes are inclined to ensure positive drainage of oil. They must avoid loops which could fill with oil.

15. The oil mist detector should be tested every day and the sensitivity checked. Lenses and mirrors should be checked periodically.

16. The two duties of the crankcase relief valve cover are to secure the valve spring and act as a deflector to direct gas to where it can do least damage.

17. The free area of the gauze firetrap must at least be equal to the area of the open valve.

18. The minimum combined area of the crankcase relief valve is not to be less than 115 cm^2 per cubic metre of crankcase volume.

19. The three sides of the fire triangle are air, fuel and the source of ignition.

Engineering Pocket Book – Answers

20. Ignition of a scavenge fire can be caused by unburned oil and carbon which has been blown from the cylinder into the scavenge spaces, this may include unburned fuel or cylinder lub oil. This may be due to incorrect combustion caused by a defective injector, faulty fuel pump or incorrect timing, lack of scavenge air, partially blocked exhaust port, low compression pressure, afterburning, by operating the engine in an overloaded condition, defective piston rings, worn liner or excessive cylinder lubrication.

21. The other danger that a scavenge fire can cause is a crankcase explosion.

22. The turning gear must be engaged to prevent seizure if the engine is stopped because of a scavenge fire.

23. The type of fire extinguishers used in fighting a scavenge fire are CO_2, dry powder, or steam.

24. If the scavenge trunking is opened too early, this may allow an ingress of air which may cause an explosion.

25. To stop the natural flow of air around the engine, canvas covers are wrapped around the turbo charger air filters to limit this.

26. After a scavenge fire, the scavenge ports should be cleaned and the trunking, together with cylinder liner and water seals, piston, piston rings, piston skirts, piston rod and gland, must be inspected.

27. The two possible causes of starting air system explosions are the continuous leaking of a defective cylinder non-return valve while the engine is running or such a valve sticking in the open position during manoeuvring.

28. Flame traps, relief valves, bursting discs or caps are fitted to the manifold of each start air valve to minimise the effects of a start air system explosion.

29. If a leaking air start valve is suspected, the engine should be stopped at the first opportunity and the valve replaced. As a temporary measure, a blank flange may be fitted to the air manifold connection to isolate the valve. But as this valve is inoperable, the engine may stop in a position from where it is unable to start. The bridge must be informed of this.

30. When the air start system is not in use, it should be shut down and all drains opened.

31. Overheating of compressor discharge air can be caused by failure of the compressor's intercooler and circulating water.

32. The pressure at which a cylinder safety valve should lift is no more than 20% above engine designed pressure.

33. The engine is turned with indicator cocks open before starting to expel any leakage, for example from the cooling system or leaking injector.

34. The three alarms that are crucial for an engine are low lub oil pressure, high cooling water temperature and low cooling water level.

35. The two sub-systems of an alarm monitoring system are simple make break switches to sophisticated sequential monitoring scanning system.

36. If there is a failure of a sensor or broken cable, the system is self-monitoring and will cause the system to give an alarm.

37. Different alarms have different indicators, eg the fire alarm will be different from the engine room alarm.

38. The recording device must be of high speed because there may be an electrical blackout where the system would have to cope with dozens of alarms in a very short period of time. If the device is not up to speed, it will not be accurate and it will be difficult to identify the correct order of events.

39. A safety system is a system that reduces dangers and risk of injury to personnel and damage to machinery. Typical safety systems are machinery auto start and reduction of power.

40. Seven parameters that will give reduction of power on an engine are:

- High scavenge temperature
- high oil mist reading
- low piston cooling pressure or flow
- high piston cooling temperature
- low jacket water pressure
- high jacket water temperature
- high exhaust gas temperature.

41. The type of system known as a second stage protection device is normally a shutdown device. This system must be independent of the first stage.

42. The dangers caused by bilge water are:

- It can be a fire hazard if there is oil in it and it flows over tank tops
- danger of free surface effect
- possibility of water damage to electrical cables and motors from splashing if over tank tops.

43. The fire detection indicator is fitted in a position where fire from a machinery space will not make it inoperative. They are usually fitted on the bridge or in a special control centre.

44. A control system on failure should be designed not to fall into an unsafe position or condition.

18 BOILERS

18.1 Answers – Boiler Types

1. The two basic types of boiler are the water tube and the fire tube.

2. The type used for high pressure, high temperature, high capacity applications is the water tube.

3. The main boiler design depends on its application, ie for motor ships, a small fire tube boiler would be satisfactory.

4. The features that are embodied in modern propulsion duty boilers are:

    - Large furnaces
    - furnaces completely water walled
    - roof firing
    - super heaters in lower temperature gas zones
    - improved methods of superheat control
    - improved soot blowing arrangements.

5. In the early D-type boiler, the superheater was fitted between the drums.

6. In modern boilers, there are normally two superheaters fitted.

7. Reheat boilers are used with reheat arranged turbine systems. Steam after expansion, ie the high pressure turbine, is returned to a reheater in the boiler. Here the steam energy content is raised before it is supplied to the LP turbine.

8. A double evaporation boiler system consists of two parts, a high pressure and a low pressure portion. The HP side is operated on a closed cycle on the waterside. Once filled with high quality water, it only needs topping up to replace any

slight leaks. The steam produced is led to the LP side, which is supported above the fired boiler. It consists of a pressure vessel containing a tube bundle through which the steam generated by the HP side is passed. The heat given up by this generates steam at a lower pressure from the water surrounding the tube bundle within the LP vessel. The LP steam is used to supply auxiliary services.

9. The other names of firetube boilers are tank boiler, smoke tube or donkey boiler.

10. Most fire tube boilers are now supplied as a completely packaged unit, which includes oil burner, fuel pump, FD fan, fuel pumps and auto controls.

11. The steam temperature at 7 bar is 170°C.

12. The temperature difference across a waste heat unit is normally around 16°C.

13. The gas temperature outlet in an exhaust gas heat exchanger is kept above 180°C to prevent low temperature corrosion occurring. There is about 10% water vapour in exhaust gas and also sulphur products and this can lead to suphuric acid forming which has a dew point of 140°C.

14. The amount of exhaust gas energy used in waste heat recovering systems is about 5 – 10 %.

18.2 Answers – Evaporators

1. The most economically viable way of making fresh water from seawater is by the evaporation process.

2. Brine is left in the heat exchanger, after vapour is taken off.

3. When scale forms on heat exchanger tubes, it reduces the efficiency of heat transfer.

Engineering Pocket Book – Answers

4. The maximum density of brine should be 64000 ppm. Above this level, a large amount of scale build-up occurs.

5. Scale deposits are removed by thermal shock when the evaporator is shut down.

6. In making potable water, it is important that the feed is not coming from a chemically treated seawater line.

7. The evaporator shell is protected from corrosion by a bonded rubber coating.

8. The materials used in the heat exchanger section are aluminium brass tubes expanded into tube plates made from admiralty brass.

9. Materials used for the demister mesh are layers of knitted monel metal wire or, alternatively, polypropylene mesh may be fitted.

10. If the distillate is above required density, the salinometer probe provides a signal which stops the distillate pump. This allows unacceptable distillate to pass over the double loop to re-enter the evaporator feed line for re-distillation.

11. The ratio of feed water to distillate is 2.75:1.

12. To prevent scale forming in the brine pipelines, a continuous blow down of brine is required which is achieved by a water-operated ejector.

13. Potable water requires further treatment, which involves the use of suitable filters and sterilising equipment.

14. The maximum period between cleanings of evaporators is at least every six months or more if required.

15. The recommended feed water temperature in flash evaporators is 80°C.

16. The gain ratio of an evaporator is the ratio of distillate output compared to heat input.

17. The type of multiple effect evaporation preferred for marine use is normally the double effect flash evaporator.

18. The type of evaporator that gives ratios of 8 or 9 is the vapour recompression plant.

19. The type of mechanical compressor preferred for this type of plant is the slow-speed blower type (rotary lobe type blower).

20. The feed of a vapour compression evaporator must have some sort of chemical added to it as evaporisation takes place at atmospheric pressure.

18.3 Answers – Boiler Mountings

1. There are two safety valves required on a boiler.

2. The safety valves are fitted to the steam drum.

3. A valve must lift a quarter of its bore to provide full flow.

4. The maximum allowable accumulation of pressure is 10% of the maximum working pressure of the boiler.

5. A conventional safety valve is adjusted by means of compression nut, screwing down onto the top of the spring plate. A cap is then placed over the compression nut and a cotter pin is placed through the valve spindle above the cap and padlocked to prevent tampering by unauthorised persons.

6. The purpose of the safety valve easing gear is to lift the safety valve and release boiler pressure in an emergency.

7. Feathering is where a thin film of steam blows across between valve faces and this will quickly cut the valve faces and lead to loss of steam.

Engineering Pocket Book – Answers

8. The full bore safety valve is designed to avoid feathering of valves.

9. The full bore safety valve consists of a main and pilot control valve, both being in direct communication with the boiler pressure. As the control valve lifts due to high pressure, it operates a piston attached to the main safety valve which causes it to open one quarter of its diameter, so giving full flow conditions for escape of boiler steam.

10. The increase in throughput with a full bore valve compared to an ordinary valve is four times.

11. Full bore safety valves solve high temperature problems because the main valve has no heavy spring to be affected by temperature.

12. The improved high lift safety valve operates depending on the design, ie one is the specially shaped vent, the other being to use the lower spring carrier in the fashion of a piston, which acted upon by the pressure of waste steam helps to compress the spring giving full blow off.

13. A high lift safety valve reduces safety valve blow down by making use of the guide sleeve to do this. The throttling effect is obtained by adjusting the vertical position of the guide sleeve, effectively controlling the speed at which the valve closes and so the amount of blow down.

14. Water in the waste steam pipe can cause a head of water to form over the valve lid so increasing blow off pressure. Also, in cold conditions, this water could freeze in the upper parts of the pipe which could lead to disastrous results.

15. The kind of valve used for a boiler stop valve is a screw down non-return valve.

Engineering Pocket Book – Answers

16. The difference between the main and auxiliary stop valves is that the auxiliary is a smaller version of the main stop, used for isolating the boiler from its auxiliary lines.

17. The purpose of the main stop valve is to isolate the boiler from the steam line.

18. Non-return valves are fitted as feed checks to prevent, in the loss of feed pressure, the boiler water flowing back into the feed line.

19. The main feed check is mounted directly on the steam drum.

20. A feed check must indicate if the valve is opened or closed.

21. If the boiler water level is too high, priming can occur with the resultant carry over of water and dissolved solid into the system.

22. Classification Societies demand feed water regulators on water tube boilers because of their high evaporation rate and small reserve of water and control is critical.

23. The usual arrangement for level indicators is that at least two are fitted to each boiler, which are mounted on the steam drum.

24. If the level disappears out of the gauge glass, immediate action must be taken to have the boiler taken out of service.

25. The fuel oil cut off must be manually reset because if it were to reset automatically, it could result in large amounts of fuel oil spraying into a hot furnace, leading to a possible explosion.

26. Deposits are prevented from forming around the needle valve of the low level device by means of an external steam line to the needle valve.

27. The purpose of the blow down valve is to enable water to be blown from the boiler in order to reduce density or, when the boiler is shut down, can be used to drain it.

28. Scum valves are fitted to the boiler to remove scum or oil from the surface of the water in the drum.

29. The purpose of the air vent is to release air from the drum or headers, either when filling the boiler or raising steam.

30. The superheater circulating valve is closed when enough demand for superheated steam is required.

31. Coolers are fitted after the salinometer valves (test cocks) to prevent flash off taking place as the pressure of the sample is reduced to atmospheric.

32. The most common type of on-load cleaning is soot blowing.

33. If on-load cleaning is not fitted, the boiler has to be shut down to clean the tubes.

34. Using steam for on-load cleaning can be expensive since all fresh water on the ship has to be made, and the make up water for the boiler must be further treated with chemicals.

35. The simplest method of steam drying fitted in the boiler drum is by fitting simple perforated plates immediately above the normal working level of the water in the steam drum.

18.4 Answers – Water Gauges

1. The limit for using test cocks as a water level indicator on a boiler is of up to 8.2 bar or with a diameter below 1.8 m.

2. The tubular gauge glass is held and sealed by nuts tightened onto soft tapered sealing rings.

Engineering Pocket Book – Answers

3. To prevent water leakage if a gauge glass breaks, a ball valve is fitted to the lower end of the gauge in order to shut off the water.

4. Plate glass guards are fitted to tubular gauge glasses to prevent injury if the glass tube shatters, especially when blowing down.

5. A diagonally striped board can be placed behind the glass so the water level is seen clearly by means of refraction causing the stripes to bend.

6. The steam side of the gauge glass is blown through first.

7. The water level in the gauge glass is caused to rise if the water cock is blocked due to condensation in the steam space gradually filling the glass.

8. When refitting the cocks in a gauge glass, it must be ensured that cocks are facing down when in the open position.

9. If the tubular glass is too short it may result in packing sleeves being squeezed over the ends of the glass, so blocking the opening.

10. A build-up of deposits in a gauge glass cock can be caused by boiler water evaporating away due to leakage.

11. Additional care must be taken if the gauge glass is indirectly mounted when blowing down as the isolating cocks on the boiler can become blocked.

12. After blowing down, the water side of the glass is open first.

13. If the water level falls to the bottom of the gauge glass, feed water should be added or supply increased immediately.

14. A gauge glass must always be tested when there is any doubt about the drum water level.

15. A reflex gauge glass works on the principle that refraction of light on the back of the rib glass plate causes the light rays to be reflected back from the steam space and absorbed in the water space. This gives a silvery appearance to the water and dark to the steam space.

16. Reflex gauge glasses are operated by chains or rods to prevent injury in the event of the glass shattering.

17. At high pressure, hot water erodes away the glass.

18. To protect the glass if the pressure is above 34 bar, a sheet of mica is placed between them to protect against erosion.

19. When overhauling a high pressure gauge glass, care must be taken in assembly to prevent undue stress being set up which will cause the glass plate to shatter when put into service.

20. The slots in a louver plate are angled upwards in a gauge glass to direct light rays in electric lamps in such a way that the actual water level in the glass appears as a plane of light when viewed from below.

21. The top and bottom of a double plate glass gauge glass must be marked so that the water shutoff ball cock goes to the bottom.

22. If remote water level indicators are fitted on the same connections as a direct reading gauge glass, before blowing down the gauge glass, the remote indicator must be isolated.

18.5 Answers – Boiler Operation

1. Four hazardous conditions to which all boilers are subject are very high temperature, abrasive and chemically aggressive constituents and the ever present risk of lax operating procedures.

Engineering Pocket Book – Answers

2. The prime consideration in boiler operation is cleanliness of both gas and water sides.

3. The major heat loss in boiler operations is the products of combustion exhausted to atmosphere via the funnel.

4. Atmospheric heat loss can be kept low by keeping heating surfaces clean and not supplying too much excess air, so that the temperature and quantity of the exhaust gases are as low as possible.

5. For maximum efficiency and perfect combustion, the air supply must be closely controlled.

6. The main source of heat loss when there is insufficient air for combustion is loss due to the presence of inert nitrogen.

7. The Funnel exhaust state that is considered good practice is where a light brown haze is obtained.

8. Unburned losses in a boiler are kept to a small value by proper maintenance of combustion equipment.

9. Fuel oil in the settling tank is heated to a suitable pumping temperature and allows water and other residue to settle out.

10. Burner connections from the fuel main are kept as short as possible to minimise the amount of cold fuel oil injected into the furnace.

11. There must be no dead-legs in the fuel main, as this will allow sludge to build up.

12. An essential safeguard for combustion equipment is regular cleaning and inspection of atomiser tips.

13. When wear is detected on the burner tip, they must be discarded and changed for new.

Engineering Pocket Book – Answers

14. The boiler air trunking should be checked for obstructions and that it is clear, especially following maintenance.

15. The tendency for fouling in a boiler is increased by poor fuel quality.

16. To prevent problems with a boiler, soot blowing equipment must be kept in good working order.

17. The air purging system should be checked for corrosion from combustion products.

18. To ensure gas side cleanliness during a shutdown, it is water washed by very high pressures.

19. Gas side cleanliness is most essential in high temperature zones.

20. Water treatment chemicals are added to the water side to remove scale build-up.

21. The cleaning currently used in water side cleaning is chemical cleaning which involves circulating weak acids around the system.

22. Following all water side boiler cleaning operations, the water side is thoroughly flushed.

23. The purity of feed water is related to the pressure of the boiler.

24. The customary practice for firing stops (shut down) to improve circulation is to briefly blow down the lower headers.

25. Before filling a boiler, it should be checked internally for tools and anything obstructing passages, tubes should be proved clear, internal surfaces should be free from oil and scale and internal fittings should be secure.

26. The gas side should be checked that it is clean and in good order.

27. Before closing the manhole, the condition of the joints should be checked.

28. Before firing the boiler, a final check should be made that water level gauges have a level in them.

29. The final check is to see if water level gauges are open and water is showing in the glass.

30. Care is needed in the initial period of firing as this is a trial for ignition period, during which the burner should ignite and a signal be received by the flame scanner.

31. The superheat circulating valve is closed when there is a flow of steam through it.

32. When the steam pressure reaches 10 bar, nuts on all new boiler joints should be followed up.

33. The time scale from flashing up to going on line is between four and six hours.

34. The procedure should be carried out more slowly when new refractory material has been installed.

35. Modern boilers are controlled with burner management and boiler control is achieved by automatic devices.

36. The firing rate must be limited when regaining pressure after loss so as not to elevate the superheater temperature.

37. If deposits accumulate on the heat recovery surface, it can lead to a fire risk.

38. Heat accumulation of soot can be caused by poor combustion and lack of cleanliness.

39. The factors that influence the likelihood of superheater tubes overheating is poor circulation of steam.

Engineering Pocket Book – Answers

40. The free hydrogen is formed for a hydrogen fire by tube metal combining with steam producing free hydrogen.

41. A hydrogen fire is extinguished by allowing itself to burn out.

42. A bypassed economiser needs soot blowing as if in service to stop build-up of deposits, which can be a fire hazard.

43. If running with bypassed heat recovery units, the inlet and outlet gas temperatures of the units should be well monitored.

44. The only cure for soot and hydrogen fires is prevention.

45. A minor furnace explosion is called a blowback.

46. The operation carried out to prevent blowbacks or furnace explosions is to purge them well before firing.

47. To prepare a boiler for an extended lay-up, completely drain and dry the boiler out using hot air, hermetically sealing it. Trays of a drying agent such as silica gel should be inserted before sealing.

18.6 Answers – Tube Failures

1. Warning of a tube failure is given during inspection, by looking for pitting.

2. The difference between tangent wall and monowall is that tangent wall tubes are situated closely together and tubes are backed by refractory. Monowall tubes have a steal strip welded between them to form a completely gas type enclosure. Only a layer of insulation is required on the outside of the construction.

3. If a membrane wall tube is plugged, the whole length of tube must be shielded from the furnace heat by means of thick shield plastic refractory.

4. The kind of repair that should be carried on a membrane wall tube is to insert a new section.